GENETIC DIVERSITY AND VARIABILITY OF HEPATITIS B VIRUS

GENETIC DIVERSITY AND VARIABILITY OF HEPATITIS B VIRUS

VERÓNICA L. MATHET
AND
MARÍA L. CUESTAS

Nova Science Publishers, Inc.
New York

Copyright © 2009 by Nova Science Publishers, Inc.

All rights reserved. No part of this book may be reproduced, stored in a retrieval system or transmitted in any form or by any means: electronic, electrostatic, magnetic, tape, mechanical photocopying, recording or otherwise without the written permission of the Publisher.

For permission to use material from this book please contact us:
Telephone 631-231-7269; Fax 631-231-8175   Web Site: http://www.novapublishers.com

NOTICE TO THE READER

The Publisher has taken reasonable care in the preparation of this book, but makes no expressed or implied warranty of any kind and assumes no responsibility for any errors or omissions. No liability is assumed for incidental or consequential damages in connection with or arising out of information contained in this book. The Publisher shall not be liable for any special, consequential, or exemplary damages resulting, in whole or in part, from the readers' use of, or reliance upon, this material.

Independent verification should be sought for any data, advice or recommendations contained in this book. In addition, no responsibility is assumed by the publisher for any injury and/or damage to persons or property arising from any methods, products, instructions, ideas or otherwise contained in this publication.

This publication is designed to provide accurate and authoritative information with regard to the subject matter covered herein. It is sold with the clear understanding that the Publisher is not engaged in rendering legal or any other professional services. If legal or any other expert assistance is required, the services of a competent person should be sought. FROM A DECLARATION OF PARTICIPANTS JOINTLY ADOPTED BY A COMMITTEE OF THE AMERICAN BAR ASSOCIATION AND A COMMITTEE OF PUBLISHERS.

Library of Congress Cataloging-in-Publication Data

Mathet, Verónica L.
  Genetic diversity & variability of hepatitis B virus (HBV) / Verónica L. Mathet, María L. Cuestas (authors).
       p. ; cm.
    Includes index.
    ISBN 978-1-60456-888-2 (softcover)
  1. Hepatitis B virus--Genetics. I. Cuestas, María L. II. Title. III. Title: Genetic diversity and variability of hepatitis B virus (HBV).
    [DNLM: 1. Hepatitis B virus--genetics--Caribbean Region. 2. Hepatitis B virus--genetics--Latin America. 3. Hepatitis B--epidemiology--Caribbean Region. 4. Hepatitis B--epidemiology--Latin America. 5. Hepatitis B--virology--Caribbean Region. 6. Hepatitis B--virology--Latin America. 7. Variation (Genetics)--Caribbean Region. 8. Variation (Genetics)--Latin America. QW 170 M427g 2008]
    QR201.H46M38 2008
    616.3'623042--dc22
                                                                                        2008025353

Published by Nova Science Publishers, Inc. ♦ New York

Contents

Preface		**vii**
Chapter 1	Introduction	1
Chapter 2	HBV Epidemiology in Latin America and the Caribbean (LAC) Region	31
Chapter 3	HBV Variants and Mutants and Its Impact on the LAC Region	53
Chapter 4	Conclusions	75
Acknowledgments		77
References		78
Index		103

Preface

In spite of the progress made in vaccine and antiviral therapy development, hepatitis B virus (HBV) infection still remains a major health care problem. More than 350 million people are chronically infected worldwide, showing differences in the severity of liver disease, clinical outcome and response to immune and antiviral-therapy. Parameters associated to the host immune system (HBV specific T and/or B-cell repertoires, defective antigen presentation and diminished Th1/Th2 response ratio) and viral factors, such as the HBV genotypes and their evolving variants, have largely contributed to the explanation of such differences.

The unique genomic structure and replication cycle of HBV contribute to the occurrence of mutations in any of its genes undergoing selection pressures. The selection of one mutant over the others warrants a biological advantage to the prevalent mutation during the replication cycle of the virus or a selective advantage to the mutant over wild-type virus in host-virus interactions as we observed in Argentina where [some novel] mutated HBsAg were detected, even in the presence of specific anti-HBs antibodies.

Reports from USA, as well as from several European and Asian countries have focused on mutations within the HBV genome that may be associated with the hepatitis B vaccine and/or hepatitis B immune globulin (HBIG) failure, diagnostic escape mutants, antiviral therapy resistance, and differential outcomes in liver disease. In contrast, studies regarding these topics are almost lacking in Latin America. Within this context, it should be emphasized that the already known mutants circulating in other parts of the world could not necessarily be mirroring those strains circulating in this part of the American continent.

Emerging data is filling the gap in our knowledge of HBV genotypes in several Latin American countries. While genotypes F and H of HBV are considered to be indigenous in this continent, genotypes A and D might be a mere reflection of a past European migration, and genotypes B and C could represent a consequence of a recent Asian migration. Few years ago, genotype G was detected in Mexico as previously identified in the USA and France. Unexpectedly, some strains ascribed to HBV genotype E (widely considered an African restricted genotype) have been observed in Argentina, even though genotypes F, A and D appear to be the most prevalent in this country.

Co-circulation of all the already known genotypes and certain HBV variants in Latin America offers a unique opportunity to study basic viral and clinical features of this infection.

Chapter 1

Introduction

"Nothing seems quite so dramatic as the unexpected eureka moment, when, escorted by the gods of good fortune, scientists somehow stumble upon answers to questions they never knew to ask. Baruch Blumberg, a U.S. geneticist and biochemist, won the Nobel Prize in 1976 after finding a virus he was never looking for..." (Carolyn Abraham, Toronto Globe and Mail).

Blumberg's discovery of the hepatitis B virus (HBV) in 1967 is considered one of the greatest medical achievements of the 20th century. The story started in the early 1960s when he was examining blood samples from diverse populations in an attempt to determine why the members of different ethnic and national groups widely varied in their responses and susceptibility to disease. In 1963 he discovered a mysterious protein in the serum of an Australian aboriginal, which he later (1967) determined to be part of a virus that caused hepatitis. The discovery of this so-called Australian antigen, later known to be the surface antigen of hepatitis B virus (HBsAg), made it feasible to screen blood donors for possible hepatitis B transmission. Further research indicated that the humoral immune response elicited to the Australian antigen conferred full protection against hepatitis B. In 1969 he developed a blood test to detect the virus and was involved in the development of the first hepatitis B vaccine.

HBV is the prototype member of the *Hepadnaviridae* family (hepatotropic DNA virus), which includes viruses recovered from a variety of animal species [Schaefer S., 2007].

Three modes of HBV transmission (perinatal, sexual and parenteral/percutaneous) have been recognized among humans, its only known natural host. Blood is the most important vehicle for transmission, but semen and saliva have also been involved [Bancroft et al., 1977; Scott et al., 1980].

It was estimated that approximately 2 billion people around the world have serological evidence of a past or ongoing HBV infection, more than 350 million are chronic carriers [WHO, 2000], and up to 40% of infected patients might develop cirrhosis, liver failure or hepatocellular carcinoma (HCC) [Lok et al., 2002]. As a result of the high HBV-related morbidity and mortality (500.000 to 1.2 million people die of HBV infection every year), it is considered one of the major world health problems [Lee et al., 1997].

The worldwide prevalence of chronic HBV infection could be categorized as high, intermediate and low endemicity depending on the percentage of HBV carriers among the general population, which is related to the age at the time of infection [Margolis et al., 1991].

Table 1. Chronic HBV infection: characteristics and worldwide distribution

Characteristics	Endemicity of infection		
	High	Intermediate	Low
Percentage of HBV carriers	≥ 8%	2-7%	<2%
Past-infection prevalence	70-95%	10-60%	5-7%
Lifetime risk of infection	≥ 60%	20-60%	< 20%
Age at the time of infection	Early childhood	All age groups	According to high-risk groups
Worldwide regions	Southeast Asia, China, sub-Saharan Africa and the Amazon Basin	Southern and Eastern Europe, Middle East, Japan and part of South America	Northern and Western Europe, North America and Australia

1.1. HBV Structure

HBV genome consists of a 3.2-kb-long partially double-stranded relaxed-circular DNA molecule that contains four open reading frames (ORF): the core (C-), the surface (S-), the X- and the polymerase (P-) genes that codify for all viral proteins. As a strategy to economize genome length to be packaged within the small-sized core particles, overlapping genes are observed, either in or out-of-frame as shown in figure 1.

Base pairing of an overlapping region flanked by two direct repeats (DR1 and DR2) allows circularization of the genome with the P protein being covalently linked to the 5´-end of the minus strand. The S-ORF has three in-frame AUG codons that divide this ORF into three regions, termed Pre-S1, Pre-S2, and S, that encode for the viral envelope proteins (figure 2). The S-ORF includes the sequence assumed to codify the binding sites for the hepatocyte. The ORF C has two initiation codons that encode the structural protein of the nucleocapsid which is important for the assembly of viral particles (hepatitis B core antigen [HBcAg]) and a soluble antigen that is secreted by the infected cells (hepatitis B e antigen [HBeAg]). The ORF X encodes for the hepatitis B x antigen [HBxAg], which is believed to have transactivating activity and a variety of other regulatory functions. This protein is not incorporated into mature virus particles. Finally, the ORF P encodes for an 831-amino acid (aa) multifunctional enzyme, called P protein, polymerase or Pol, which is responsible for the following functions in viral replication: 1) encapsidation of the viral pregenomic RNA; 2) priming of DNA synthesis; 3) reverse transcription of the pregenomic RNA to the double stranded DNA; and 4) RNAse-H activity for the degradation of the pregenomic RNA [Schories et al., 2000]. This viral protein is also an ideal target for antiviral compounds used in the treatment of chronic hepatitis B.

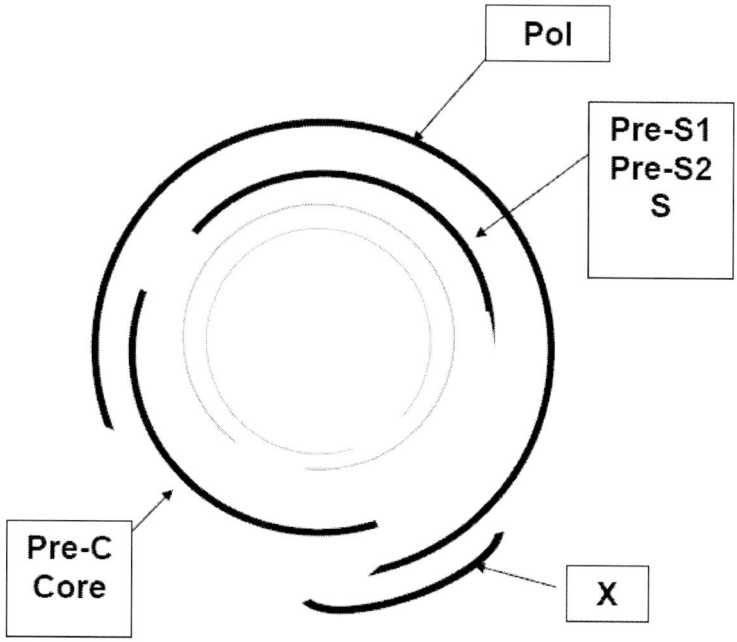

Figure 1. Schematic representation of the HBV genome (grey thin lines) and the encoded proteins. (black thick lines). A partially double stranded DNA of only approximately 3.2 kb codes for all proteins by using an overlapping (in-frame and out-of-frame) gene structure. Note that the Pol protein is encoded by a gene that overlaps the remaining three open reading frames (See the text).

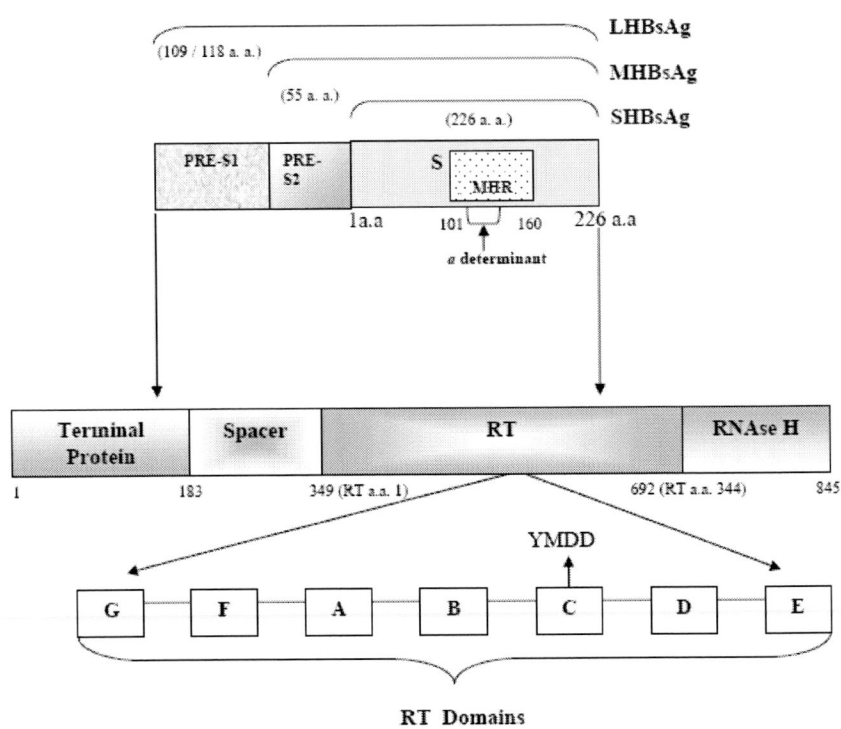

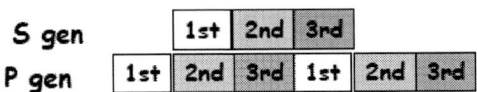

Figure 2. A representation of the overlapped S and P genes is shown. S gene (in-frame) encoded envelope proteins (large, middle and small HBsAg) are depicted. The major hydrophilic region (MHR) is shown between aa 101 and 160. The *a* determinant is placed inside the MHR. The catalytic site of the HBV reverse transcriptase (RT) is indicated with an arrow pointing to the YMDD amino acidic region of the RT C domain. The overlapped position within a given codon in the S and (out-of-frame) P genes is shown at the bottom.

The structure of the HBV polymerase consists of four conserved domains that have been determined by comparison with the HIV reverse transcriptase sequence and confirmed by genetic and functional studies [Zoulim F., 1999]. These four different domains that could be distinguished within the P protein are:

a) the terminal protein, which is involved in the priming for reverse transcription;
b) the spacer region;
c) the reverse transcriptase (RT) / polymerase domain, where the very conserved YMDD (Tyr-Met-Asp-Asp) motif is part of the catalytic site; and
d) the RNAse-H region, which removes the RNA template.

The terminal protein is the part of the polymerase that harbors the primer for reverse transcription. Therefore, it is where the synthesis of the minus strand-DNA is initiated [Kreutz C., 2002]. Point mutations in this region are reflected as the inability of the viral polymerase to encapsidate the viral pregenomic RNA, leading HBV to be unable to continue its replication.

The spacer region is dispensable for enzymatic activity and tolerates mutations, but its own function is still unknown.

The RT/polymerase region of the viral polymerase is a functional domain where reverse transcription and synthesis of the second DNA strand, take place. It can be divided into seven domains labeled A, B, C, D, E, F and G. Its catalytic site is within domain C, around the amino acidic-sequence YMDD (figure 2), a highly conserved motif among hepadnaviruses and retroviruses, which seems to be the nucleotide recognizing site of this viral enzyme. Drug-resistant mutants harbour point mutations in the C and A or B domains. Within the C domain, point mutations in the YMDD motif are the most common and are of clinical significance because many of them may be selected during antiviral therapy with nucleoside/nucleotide analogs, such as lamivudine, contributing in this way to therapy failure. Those YMDD-lamivudine-resistant mutants that changed the YMDD-motif into YIDD, YVDD, YRDD and YSDD are among them.

Finally, the RNAse-H region of the viral polymerase is another functional domain where degradation of the pregenomic RNA takes place [Kreutz C., 2002].

In addition to the well known T-cell epitopes located within C and S proteins, some aa-sequences located within the RT and the RNAse-H domains of the HBV P protein are highly conserved, being CD8+/CD4+ T-cell epitopes that might contribute to both the viral clearance and the development of subclinical forms of hepatitis B disease.

Regulatory elements and the C, P, S and X promoters of the hepatitis B virus are located within the protein encoding regions. Two enhancers (enh I and enh II), one silencer element (SE), one encapsidation signal (epsilon [ε]) and one glucocorticoid responsive element (GRE), have been described.

The mean half-life of serum HBV DNA is estimated at 1 to 2 days and the rate of HBV virion production may be as high as 10^{11} virions per day [Wai et al., 2004].

Electron microscopy of partially purified preparations of HBV isolated from the plasma of individuals who are either undergoing an acute infection or have become carriers of hepatitis B virus, shows three types of particles. The first of these forms is a double-shelled particle referred as the Dane particle, after its discoverer D. S. Dane. [Dane et al., 1970]. This particle constitutes the complete virion. It is 42 nm in diameter and contains a 28-nm core structure designated as the hepatitis B core antigen (HBcAg) that is surrounded by an outer lipoprotein envelope that contains the three related envelope glycoproteins (Pre-S1, Pre-S2 and S). Recently, it was elegantly shown that the virions contain capsids with either $T = 3$ or $T = 4$ icosahedral symmetry. Projections extending from the lipid envelope were attributed to surface glycoproteins. Their packing was unexpectedly non-icosahedral but conformed to an ordered lattice (figure 3). These structural features distinguish HBV from other enveloped viruses [Dryden et al., 2006]. Viral DNA and viral polymerase are within the core. The second of these forms is a spherical particle averaging approximately 20 nm in diameter. These particles are the most predominant form in serum, usually present in a 10,000- to 1,000,000-fold excess over Dane particles. Finally, smaller quantities of a heterogeneous population of filamentous particles, all with a diameter of 20 nm and variable length (up to approximately 350 nm) have been identified.

The 20-nm spheres and filaments are empty particles, made up exclusively of envelope proteins, host-derived lipids (approximately 30% by weight) and N-linked carbohydrates. Main lipids include phospholipids, cholesterol esters, and triglycerides. These particles lack nucleic acid altogether and are hence non-infectious. The 22-nm spherical form contains approximately 89% of S protein, 10% of Pre-S2 protein, and 1% of Pre-S1 protein. The filaments consist of the same ratio of proteins as the infectious Dane particles, comprising approximately 70% of S protein, 10% of Pre-S2 protein, and 20% of Pre-S1 protein [Lee and Locarnini, 2004]. In pure form empty particles are highly immunogenic and efficiently induce a neutralizing anti-HBs antibody response. In fact, the 20-nm spheres, purified from the serum of hepatitis B chronic carriers, served as the initial form of HBV vaccine prior to the development of recombinant HBsAg preparations. Natural infection thus presents the seeming paradox of efficient progression, despite the accompanying production of highly immunogenic particles that can elicit neutralizing host responses [Knipe et al., 2001].

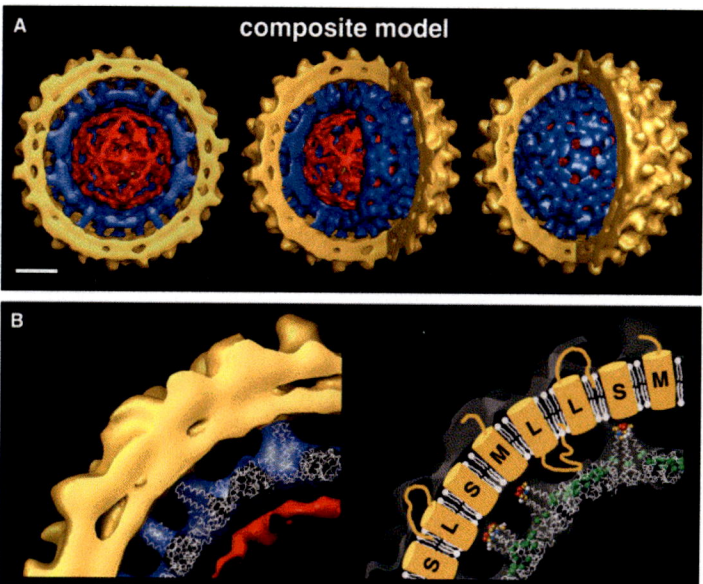

Figure 3. Model and Cartoon of HBV Virions reprinted from Mol. Cell 22 (6): 843-50, Dryden KA *et al*, "Native hepatitis B virions and capsids visualized by electron cryomicroscopy", Copyright (2006), with permission from Elsevier. (A) Cut away views of a composite model of the HBV virion comprised of a T=4 icosahedral capsid with 120 spikes and an outer envelope with protein projections spaced 60 Å apart. Views are cross-sections (left), and two cutaways. (B) X-ray crystal structure of recombinant capsid docked into the cryo-EM density map of the virion capsid (left). The tips of the core spikes are in close apposition but do not penetrate the envelope. Additional details and cartoon of interpretation (right). The surface protein projections are ascribed to HBsAg and are designated as large (L), medium (M), and small (S) according to the three start codons of the HBsAg ORF. Residues in the capsid tip are colored according to charge and hydrophobicity (negative, red; positive, blue; hydrophobic, gold; and hydrophilic, grey) and show a mixed distribution of charge. Residues denoted by green spheres have been implicated in viral envelopment via interaction with the L protein. HBcAg (yellow) modeled as S box with M and L loops. Note that ~50% of the L molecules have an interior loop that is predicted to be disordered, which interacts with specific residues in HBcAg (green spheres). Bar, 100 Å.

As mentioned above, the HBV envelope is a complex structure composed exclusively of host-derived lipids and three related viral proteins: the large (LHBsAg), the middle (MHBsAg), and the small (SHBsAg) envelope proteins, which are also known as Pre-S1, Pre-S2 and S, respectively. Among them, the small S protein, often called HBV surface antigen (HBsAg) is the major component of the viral envelope. About half of the S protein is in the non-glycosylated form (p24) and the other half is in the N-glycosylated form (gp27) [Antoni et al., 1994]. These two forms present identical amino acidic sequences and only differ by the presence of a complex oligosaccharide chain with a sialic acid component attached to gp27 at Asn 146 due to the presence of the consensus glycosylation sequence Asn-X-Thr/Ser in that position. It is interesting to note that in spite of the presence of two other potential glycosylation sites, one at Asn-3 and the other at Asn-59, these positions are not glycosylated. A possible explanation could be the failure of these sites to enter the lumen of the endoplasmic reticulum (ER) where glycosylation takes place, or the assumption that this part of the protein has an unfavorable conformation for glycosylation [Eble et al., 1987].

The other two envelope proteins, Pre-S1 and Pre-S2, also exist in two forms, either glycosylated at Asn 146 of the S sequence or non-glycosylated at this site. The fact that Asn 146 is found in either form, (i.e .glycosylated or non-glycosylated) might be explained by the observation that the S protein seems to be unique in adopting two distinct conformational states in co-translational translocation in the ER membrane [Huovila et al., 1992]. In addition, the Pre-S2 protein is also glycosylated at Asn 4 [Prange et al., 1995].

The three HBV envelope proteins are translated from distinct mRNA. The S protein is coded for by the S gene and is made up of 226 aa. The Pre-S2 protein is coded for by the Pre-S2 region (55 aa) and the S gene, while the Pre-S1 protein is coded for by the Pre-S1 region (108 or 119 aa, depending on genotypes), the Pre-S2 region, and the S gene. Thus, all these three envelope proteins share the same carboxyl terminus, i.e., HBsAg [Yamamoto et al., 1994] and are easily detected by conventional immunoassays used for diagnosing HBV infection and screening blood donors around the world. It was suggested that the Pre-S1 and possibly the Pre-S2 protein, might mediate the binding and entry of the virus into the hepatocytes. The Pre-S1 protein also regulates the nuclear pool of covalently closed circular DNA (cccDNA).

Finally, the three envelope proteins (Pre-S1, Pre-S2, and S) altogether participate in the generation and secretion of the enveloped virus. It is probable that the Pre-S2 protein may not be essential in this process [Prange et al., 1995].

HBsAg particles are pleomorphic and to date, it has not been possible to crystallize them; consequently, their exact three-dimensional structure is still unknown. However, by combining information about their hidrophobicity, hydrophobic moments, flexibility, secondary structure prediction and antigenicity, several models were proposed for the putative structure of these HBsAg-containing particles. Nevertheless, it is widely accepted that their antigenicity is mainly dependent on the conformation of the *a* determinant, placed within the major hydrophilic region (MHR). This hydrophilic region encompasses aa 101 to 160 of the S protein and is exposed on the surface of both virions and subviral particles. This region is highly immunogenic and is potentially under selective pressure of the immune system. The *a* determinant is a complex conformational region and disulfide bridges among highly conserved Cys residues are important in maintaining the native conformational structure of this major determinant of the HBsAg (figure 4). Of the 14 Cys residues present in the 226-aa-sequence common to all three HBV envelope proteins, 8 are located within the MHR of HBsAg and are important in keeping the conformation and thus the antigenicity of the *a* determinant. It is thought that the *a* determinant consists of two loops maintained by disulfide bridges between Cys107 and 138, and Cys 139 and 147. This major determinant, which is considered the main neutralization epitope, is common to all HBV genotypes, and antibodies directed to it usually confer protection against reinfection with any of the HBV genotypes. Amino acid substitutions within the *a* determinant can lead to conformational changes, which in turn can affect the binding of the neutralizing antibodies [Cuestas et al., 2006].

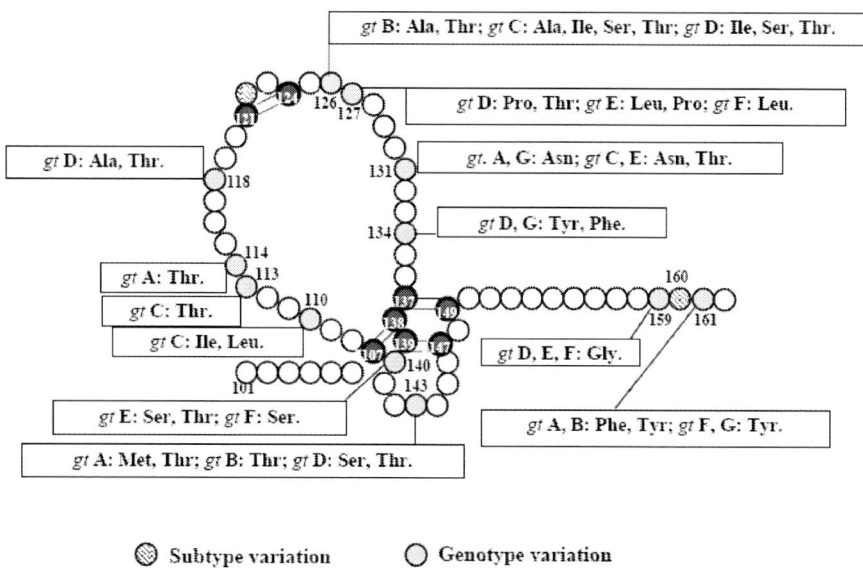

Figure 4. Schematic representation of the MHR in the S protein (aa 101 to 160). The black shadowed circles represent Cys residues. Disulfide bonds are shown by double bars. Aminoacids variation dependent on genotype (grey shadowed circles) is shown according to Osiowy [Osiowy C., 2006]. The d/y and w/r subdeterminants are mutually exclusive in key amino acidic positions 122 and 160.

1.2. Viral Pathogenesis

HBV may cause acute and chronic hepatitis. Acute hepatitis can either resolve spontaneously (acute self-limited hepatitis), lead to liver failure (fulminant hepatitis) or become chronic. Acute hepatitis develops mainly after horizontal transmission. In contrast, chronic hepatitis develops mainly after vertical transmission. Patients with long-standing active liver disease are prone to develop end-stage liver disease and HCC [Burda et al., 2001]. It is now recognized that the natural course of chronic HBV infection consists of four phases: 1) immune tolerance, 2) immune clearance (HBeAg positive-chronic hepatitis B), 3) inactive carrier state, and 4) reactivation (HBeAg negative-chronic hepatitis B) [Yim and Lok, 2006]. HBeAg has been undoubtedly demonstrated to act as an immune regulator since it may predispose to viral persistence during perinatal infections as well as prevent severe liver damage during adult infections [Chen et al., 2005]. This discovery underscores the relevance of those HBV mutants lacking HBeAg expression, as it will be discussed within Section 3.2.

Host, environmental and viral factors have been involved in the variable natural history of acute and chronic HBV infection. Age at time of acute infection is the strongest host factor associated with developing chronic HBV; infants and children are at a markedly increased risk of developing chronic HBV infection as compared with adults. In this regard, while 95% of infected neonates with immature immune systems become asymptomatic chronic HBV carriers, only 30% of children infected after the neonatal period but under 6 years of age, and 3-5% of adults do so. The remainder has acute infections resulting in viral clearance. Host immune response and host immune status are also important. The severity of the hepatocyte injury in an HBV infected patient reflects the vigor of the immune response, as the virus in most circumstances is not cytopathic: the more complete the cellular immune response is, the greater likelihood of viral clearance and more severe liver injury the patient has [reviewed in Rehermann and Nascimbeni, 2005]. It has also been established that a Toll like receptor signaling is also involved in the *in vivo* inhibition of HBV in non-parenchimal cells such as dendritic cells [Isogawa et al., 2005].

It has been recognized that HBV clearance from acutely infected hepatocytes is due to a vigorous, polyclonal and muti-specific cellular immune response. According to recent results obtained from experimentally infected chimpanzees, both cytopathic and non-cytopathic mechanisms of viral clearance are involved [Guidotti et al., 1999]. Importantly, cytokines such as IFN-γ (gamma-interferon) and TNF-α (Tumor Necrosis Factor alfa), appear to be involved in this non-cytopathic viral clearance. This novel anti-viral mechanism was observed for the first time when HBV infections were studied. When simulation studies were carried out with this model, the best fit indicated that during the early phase of HBV clearance, non-cytopathic T-cell effector mechanisms inhibited viral replication and greatly shortened the half-life of the long lived cccDNA transcriptional template, thus limiting the extent to which cytopathic T-cell effector functions while hepatocytes destruction is required to terminate acute HBV infection [Murray et al., 2005].

Immunosuppressed and/or HIV-coinfected individuals show an increased risk of developing chronic HBV as well as HBV-related morbidity and mortality. In fact, a direct cytopathic effect was observed when severe combined immune deficient mice (uPA-SCID), harbouring human liver cells, were infected with HBV, and ground-glass lesions, cell damage as well as cell death were documented, resembling those lesions observed in severely immunocompromised humans [Meuleman et al., 2006]. Environmental factors such as alcohol consumption and hepatitis C virus (HCV) and/or hepatitis D virus (HDV) coinfection have also been implicated in disease progression. The role of viral factors, including genotypes, mutations, and replication in the natural history of chronic HBV, is far from being fully understood [Wai et al., 2004; Yim and Lok, 2006].

The HBV life cycle within its host may be split in two phases (replicative and integrative), each one divided into 2 stages. The first stage is characterized by immune tolerance with the detection of HBsAg after an incubation period of 4 to 10 weeks, shortly followed by antibodies against the HBV core antigen (anti-HBc antibodies). The presence of HBeAg and large quantities of HBV DNA in serum signal the period of active viral replication.

In the second stage, an immunologic response develops or improves, leading to cytokine stimulation, direct cells lysis and the inflammatory process. Once this response is under way, aspartate and alanine aminotransferase (AST and ALT) levels increase and HBV DNA titers in blood and liver begin to drop.

When the host is able to mount a response that eliminates infected cells or greatly diminishes their number, active viral replication ends and the third stage begins. In this stage, HBeAg is no longer present, and antibodies to HBeAg become detectable. The infection has been cleared with a marked decrease in viral DNA (although many patients remain positive for HBV DNA, as detected by PCR) and ALT levels become normal. However, in some chronic cases, the S gene may be integrated into the host's hepatocyte genome, due to which patients remain positive for HBsAg. All these viral and host immune interactions are reflected in the clinical stages of the disease [Mc Mahon B.J., 2004]

In an acute self-limited HBV infection, the clearance of HBsAg and the presence of antibodies to HBsAg (anti-HBs) signal the development of full immunity to the virus and the beginning of the fourth and final stage. HBV DNA is no longer detected, and the patient is unlikely to become ill after a potential reinfection or to have a reactivated infection, since anti-HBs antibodies confer protective immunity [Lee et al., 1997].

On the other hand, in a persistent HBV infection, the first three stages unfold as in a self-limited infection, but HBsAg remains in the blood and virus production continues, often for life. However, in most people with chronic infection, levels of viremia are lower than during an acute infection, especially in those with anti-HBe antibodies. High titers of HBV in the blood are usually indicative of the continuous presence of HBeAg [Ganem et al., 2004]. Superinfection of HBV on hepatitis B carriers may occur when the same (homotypic) or a different (heterotypic) genotype infects the patient. Acute exacerbations may take place, albeit at a low frequency even in hyperendemic areas of HBV infection [Kao et al., 2001].

Both the clearance of HBsAg from serum and the appearance of anti-HBs antibodies have been associated with the resolution of hepatitis in acute or chronic hepatitis B infection. However, HBV DNA may persist in serum, liver and peripheral blood mononuclear cells (PBMC) long after circulating HBsAg has disappeared in both self-limited acute hepatitis and even after successful antiviral treatment of chronic HBV infection. Furthermore, HBV DNA has also been detected in serum and PBMC of hemodialysis patients with neither clinical nor biochemical symptoms of liver disease, and in the absence of circulating HBsAg [Cabrerizo et al., 2000]. HBV DNA detection in patients who lack detectable HBsAg has been defined as "occult HBV infection" [Cacciola et al., 1999]. This occult infection is mostly found in anti-HBc and/or anti-HBs positive individuals. It also occurs in individuals negative for all serologic markers of HBV infection. These patients with occult infection may carry both integrated HBV DNA and free HBV cccDNA [G. Raimondo et al., 2000]. The reasons for the absence of HBsAg detection are inhibition of expression by HCV or HDV co-infection, and/or low-level HBsAg synthesis under the limit of detection of screening assays, the presence of immune complexes and the emergence of HBsAg mutants [Weber B., 2005, 2006; Allain JP., 2006].

1.3. Viral Variability and Diversity

A "dictionary" *working definition* of "diversity" describes a range of people or things that are very different from each other. Likewise, the term "variability" is defined as the fact of something being likely to vary. However, in many recently published virological papers both terms have been indistinctly used. To the purpose of this review, the meaning of both words will be kept as "linguistically" described above.

Although HBV is a DNA virus, it replicates in infected cells at a high rate through an RNA intermediate because of the RT activity of its viral polymerase. The synthesis of this pregenomic RNA by a cellular RNA polymerase II is directed by the core promoter, and its encapsidation into the core particle requires the P protein, the encapsidation signal ε on the pregenomic RNA, and the core protein. Reverse transcription of this pregenomic RNA is performed by the P protein, which has to switch the template three times to accomplish the synthesis of the final virion HBV-DNA. The HBV polymerase lack of proofreading activity leads to a high rate of mutations, estimated at 1.4 to 3.2×10^{-5} nucleotide substitutions per site per year (nt/site/yr)[Ganem et al., 2004]. This is one to two orders of magnitude lower than other viruses that lack polymerase-associated proof-reading functions, such as RNA viruses [Brunetto et al., 1999], and it is four orders of magnitude greater than in "common" (without reverse transcriptase) DNA viruses [Pumpens et al., 2002]. As a result, HBV populations may evolve more quickly than most DNA viruses in response to environmental factors, and the emergence of viral HBV mutants under the pressure of such factors is likely to be a frequent event. Recently, after studying the evolution of 8 full length genotype B DNA genomes from HBeAg minus patients (without treatment) throughout a 25-yr period, an estimated rate of 7.9×10^{-5} nt/site/yr were recorded [Osiowy et al., 2006]. Interestingly, when such rate was analyzed from a patient infected with an HBeAg minus genotype F strain in presence of anti-HBs antibodies throughout a 3-yr period, a rate of 2.7×10^{-3} nt/site/yr was recorded [Mathet et al., 2007, López et al., 2007].

In addition to the introduction of point mutations carried out by the P protein during reverse transcription, or during pregenomic RNA synthesis by cellular RNA polymerase II, other potential mechanisms to generate variability are: editing of viral DNA by cellular cytidine deaminases, (promoting G to A and C to T substitutions) and deletions/insertions in HBV genomes by splicing of the pregenomic RNA by cellular topoisomerase I [reviewed by Günther S., 2006]. A novel type of HBV mutation which was coined as "replacement mutation" has recently been recognized for some (but not all) genotype E HBV strains [Fujiwara et al., 2005]. As discussed below, recombination events seem to have a prime significance in HBV genomic diversity. In this regard, recombinations have been documented not only among strains from the same genotype, but also from different genotypes. Moreover, some human HBV strains (genotype C) exhibit genetic prints of cross-species recombination with either chimpanzee or gibbon HBV variants [Simmonds and Midgley, 2005].

All these factors have contributed to a wide variation and, consequently, a diversification of HBV strains (figure 5). Genotypes, subgenotypes, and HBsAg subtypes represent genetically stable viral populations that share a common, separate evolutionary history (genetic diversity). They emerged in specific human populations and migrated with their hosts to other areas of the world, leading to their present geographic distributions [Echevarría et al., 2006]. Moreover, recombination between genotypes generates novel variants that contribute to the genetic diversity of HBV.

Figure 5. HBV genetic variability and HBV genetic diversity. Complementary interactions influence both of them and account for HBV evolution. Cartoon designed by Maria Mercedes Martinez.

In terms of genetic variability, a distinction between HBV genotypes and mutants has to be made. The different genotypes of HBV are (relatively) stable forms of the virus, which are the result of random changes selected over years of population pressure. In other words, the term genotype is applied to the replication competent forms into which the genomic sequence has stabilized after a prolonged period of time [Kramvis et al., 2005b]. Mutants arise in individuals under medically or naturally (chronic hepatitis B) induced immune pressure. They include vaccine and HBIG escape mutants. In liver transplant recipients, high titer and persistent administration of HBIG in a situation with low viral titer and a large number of susceptible cells (liver transplant) is an ideal breeding ground for mutants. In chronic HBV infection, naturally occurring escape mutants may be selected by the immune response of the HBV carrier [Weber B., 2005].

Natural polymorphisms of HBV or "variants" are naturally occurring variations observed within an individual and among populations compared with published wild type strains such as Pre-C and core promoter variants. It has been recommended that variants that arise under specific selection pressure such as HBIG or lamivudine treatment, which are associated with a specific phenotype, should be termed "mutants" [Wai et al., 2004].

Consensus definitions were recently published to discriminate between naturally occurring variants and mutants. A HBV variant is characterized as any naturally occurring variation from published wild-type sequences. A HBV mutant is defined as a variant that develops under specific selection pressure and that has been shown to confer a specific phenotype [de Franchis et al., 2003].

HBV mutants are stable over time and can be horizontally or vertically transmitted [Weber B., 2005].

The generation of continuously evolving variants/mutants within a given HBV population within a host contributes to the variability of a given strain.

The basic classification of genomic mutations proceeds from their nature and identifies three major categories:

- ➤ Point Mutations: nucleotide changes that affect a single base pair in one gene. There are two types of point mutations: transitions and transversions. Transitions are the most common, and consist of the substitution of one pyrimidine by another pyrimidine, or of one purine by another purine. Transversion is the substitutuion of one purine by one pyrimidine and vice versa.
- ➤ Rearrangements: nucleotide changes that affect a large region of a gene. They could be insertions of a transposable sequence within a gene, or deletions or the loss of a nucleotide sequence.
- ➤ Frameshift: is produced by insertion or deletion of a single base that changes the reading frame.

1.3.1. HBV Subtypes

The earliest demonstration of the diversity of the virus was the grouping of HBV isolates into HBsAg subtypes [Couroucé et al., 1976]. HBsAg epitopes involved in the expression of subtype specificities are situated in an external antigenic region of the molecule that includes the two loops involved in the structure of the *a* determinant, which is placed within the MHR, as observed in figures 2 and 4. The MHR is exposed on the surface of both virions and subviral particles.

Antibodies directed against the *a* determinant elicited by a particular HBsAg subtype can neutralize the infectivity of HBV particles and confer cross-protection against all subtypes, because this *a* determinant is common to all HBV subtypes, while the *d/y* and *w/r* subdeterminants are mutually exclusive (table 2 and figure 4). Key positions for subtype assignment involve aa 122 (Lys in subtype *d* or Arg in subtype *y*) and 160 (Lys in subtype *w* and Arg in subtype *r*). HBsAg subtype identification requires testing with reference panels of polyclonal antisera or monoclonal antibodies, but can also be predicted from the sequence of the viral genome region encoding the HBsAg by identifying the amino acids encoded at specific positions [Magnius et al., 1995]. Ten serological subtypes have been identified as: *ayw1, ayw2, ayw3, ayw4, ayr, adw2, adw3, adw4q⁻, adr, adrq⁻*. Notation of these subtypes is done by acrostics that begin with the common *a* determinant and are followed by the subtype determinants of that particular strain (table 2).

1.3.2. HBV Genotypes

Over the last decade, subtype determination has gradually been replaced by genotyping. By an intergroup divergence of more than 8% in the complete genome sequence of HBV strains obtained worldwide, eight different HBV genotypes, designated A-H, have been identified [Norder et al., 2004; Schaefer S., 2005 and 2007] (figure 6). These genotypes have a distinct geographic distribution. Genotype A is mainly found in Northern Europe, North America, India, and Africa. Genotypes B and C are prevalent in Asia. Genotype D is universally distributed but is the predominant genotype in the Mediterranean region, while genotype E is found mainly in West Africa. Genotype E shows a strikingly low level of diversification among already sequenced strains and, even more important, is the most divergent *a* determinant among all genotypes. The latter fact appears to be an influencing factor for diagnostic assays testing HBsAg, also responsible for vaccination failure [Karthigesu et al., 1999].

Table 2. Amino acid residues specifying HBsAg determinants

Subtype	Amino acid residue at HBsAg position						
	122	127	134	159	160	177	178
ayw1	Arg [R]	Pro [P]	Phe [F]	Ala [A]	Lys [K]	Val [V]	Pro [P]
ayw2	Arg [R]	Pro [P]	Tyr [Y]	Gly [G]	Lys [K]	Val [V]	Pro [P]
ayw3	Arg [R]	Thr [T]	Phe [F]	Gly [G]	Lys [K]	Val [V]	Pro [P]
ayw4	Arg [R]	Leu / Ile [L] / [I]	Phe [F]	Gly [G]	Lys [K]	Val [V]	Pro [P]
ayr	Arg [R]	Pro [P]	Phe [F]	Ala [A]	Arg [R]	Val [V]	Pro [P]
adw2	Lys [K]	Pro [P]	Phe [F]	Ala [A]	Lys [K]	Val [V]	Pro [P]
adw3	Lys [K]	Thr [T]	Phe [F]	Ala [A]	Lys [K]	Val [V]	Pro [P]
adw4q⁻	Lys [K]	Leu [L]	Phe [F]	Gly [G]	Lys [K]	Val [V]	Gln [Q]
adr	Lys [K]	Pro [P]	Phe [F]	Ala [A]	Arg [R]	Val [V]	Pro [P]
adrq⁻	Lys [K]	Pro [P]	Phe [F]	Val [V]	Arg [R]	Ala [A]	Pro [P]

Genotype F shows the highest divergence among the genotypes and is indigenous to aboriginal populations of the Americas. Genotype G has initially been found in France and the United States, and the newly described Amerindian genotype H has been identified in Central America. [Arauz-Ruiz et al., 2002]. Although the virus structure is essentially identical in all genotypes, they have different genomic lengths being HBV genotype G the longest. Its 3248 base pairs (bp) are attributed to an insertion of 36 bp at codon 2 of the core gene [Stuyver et al., 2000]. This HBV genotype has also two stop codons at positions 2 and 28 of the Pre-C region, either of which prohibits the translation of the HBeAg precursor [Carman et al., 1989; Okamoto et al., 1990]. In spite of this, HBeAg is commonly detected in sera from individuals infected with genotype G [Mizokami et al., 1999] due to a frequent coinfection with genotype A [Kato et al., 2002a].

However, the exclusive monoinfection with HBV genotype G (without synthesis of HBeAg) has recently been demonstrated [Chudy et al., 2006].

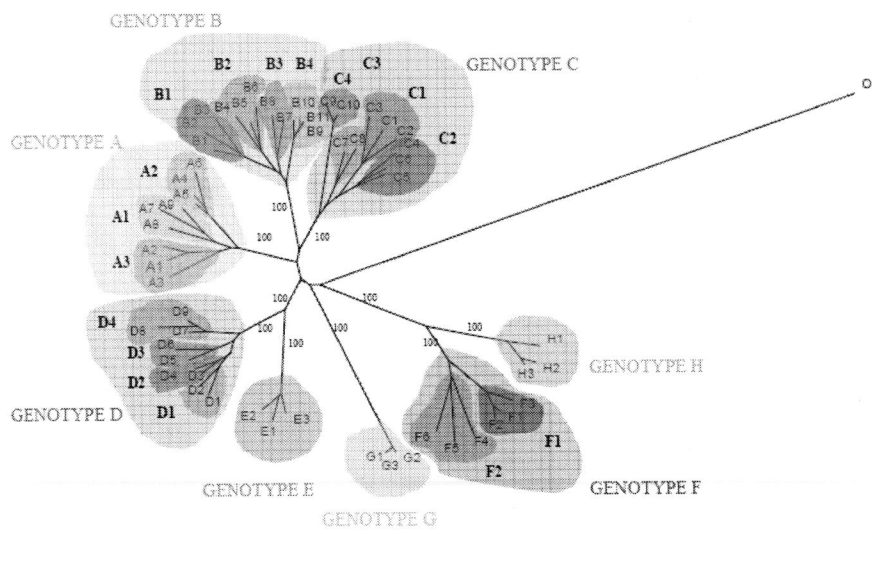

Figure 6. An unrooted phylogenetic neighbour-joining (NJ) tree of 54 full-length HBV genomic sequences representative of the eight genotypes and their corresponding subgroups. Bootstrap values are indicated in the tree roots. GenBank accesion numbers for the representative sequences included in the phylogenetic analyses are: genotype A subgenotype A3 (AB194951 [A1], AB194952 [A2], AB 194950 [A3]); genotype A subgenotype A2 (V00866 [A4], AY233280 [A5], S50225 [A6]); genotype A subgenotype A1 (AB076679 [A7], M57663 [A8], AY233282 [A9]); genotype B subgenotype B1 (D00329 [B1], AB010292 [B2], AB073854 [B3]); genotype B subgenotype B2 (X97850 [B4], AB073839 [B5], AB076679 [B6]); genotype B subgenotype B3 (AB033554[B7], AB033555 [B8], D00331 [B9]); genotype B subgenotype B4 (AB117759 [B10], AY033073 [B11]); genotype C subgenotype C1 (AB112065 [C1], AY217371 [C2], AB112348 [C3]); genotype C subgenotype C2 (AF182802 [C4], D50520 [C5], AF533983 [C6]); genotype C subgenotype C3 (X75656 [C7], X75665 [C8]); genotype C subgenotype C4 (AB048704 [C9], AB048705 [C10]); genotype D subgenotype D1 (AF418679 [D1], AY74176 [D2]); genotype D subgenotype D2 (AB205126 [D3], Z35716 [D4]); genotype D subgenotype D3 (AJ344117 [D5], X85254 [D6]); genotype D subgenotype D4 (AB033559 [D7], AB048701 [D8], AB048702 [D9]); genotype E (X75664 [E1], AB091255 [E2], AB091256 [E3]); genotype F subgenotype F1 (AY090461 [F1], AB090459 [F2], AF223963 [F3]); genotype F subgenotype F2 (AY090455 [F4], AF223965 [F5], AB036910 [F6]); genotype G (AB056513 [G1], AB056515 [G2], AB064313 [G3]); genotype H (AY090460 [H1], AY090454 [H2], AY090457 [H3]). The full-length genome of the woolly monkey hepadnavirus, the most divergent primate hepadnavirus, was included as an outgroup sequence (O: AF046996).

Table 3. Relationship between genotypes, subgenotypes, HBsAg subtypes and their main geographical distribution

HBV genotype	HBV subgenotype	HBsAg subtype	Mainly observed geographical distribution
A	A1	adw2	Africa, India, Japan
		ayw1	South Africa, Malawi, the Philippines
	A2	adw2	Europe, North America
	A3	ayw1	Cameroon, Gabon
B	B1	adw2	Japan
	B2	adw2	China, Taiwan, Hong Kong, Vietnam
		adw3	Thailand
	B3	adw2	Indonesia, Polynesia, Hawaii
	B4	ayw1	Vietnam, France
C	C1	adr	China, Korea, Japan, Thailand, Laos, Malaysia, Bangladesh
		ayr	Vietnam, Korea, Japan, Alaska
		adw2	Japan, China, Vietnam, Indonesia, Tibet
	C2	adr	China, Korea, Japan, Thailand, Laos, Costa Rica, France
		ayr	Vietnam, Korea, Japan, Alaska
	C3	adrq⁻	Oceania, Hawaii, New Zealand, Japan
		adw2	Japan, China, Vietnam, Indonesia, Tibet
	C4	ayw3	Northeastern Australia (Aborigines)
D	D1	ayw2	Mediterranean area, Middle East, India, Eastern Europe
	D2	ayw3	Europe
		ayw4	India, Japan
		adw3	West and Central Africa, USA
	D3	ayw2	Sweden, Spain
		ayw3	South Africa and Alaska
	D4	ayw2	Europe, USA, injecting drug users, Oceania and Somalia
E	−	ayw4	Africa, Argentina*
F**	F1	adw4q⁻,	Central America, Alaska
	F2	ayw4	South America, Polynesia
		adw4q⁻, ayw4	
G	−	adw2	France, USA, Mexico
H	−	adw4	Nicaragua, Mexico, USA

*To date only two cases have been documented [Mathet et al., 2006].
**See comments about F subgenotypes within the text (Section 2.3.7).

In addition, "subgenotypes", defined as subgroups of HBV genotypes with between 4% and 8% intergroup nucleotide difference throughout the complete genome, have been identified within genotypes A, B, C, D and F. There is a correlation between serological subtypes and HBV genotypes, although several subtypes are encoded by more than one genotype (table 2). HBV groups defined by each genotype-subtype association often display a characteristic geographical distribution, being these features useful in identifying the import of HBV strains into a given population (table 3). When phylogenetic differences throughout the whole genome are under 4% [Kramvis et al., 2005b], a clade is designed. Within a given sub-genotype, one or more clades may be present.

How to Assign a Given Genotype?

Determination of HBV genotype by PCR amplification and DNA sequencing. An important issue to be taken into account when both epidemiological and clinical studies are undertaken is the methodology used to define a given genotype or subgenotype (table 4) [Bartholomeusz and Schaefer, 2004]. Several procedures have been proposed, although the PCR amplification followed by full length sequencing of the viral genome is widely considered the gold standard methodology. As stated in other parts of this review, A to H genotypes have been defined based on greater than 8% nucleotide variation throughout the whole HBV genome. Although this methodology is the best tool currently used, it is expensive, time consuming, requires a considerable expertise on phylogenetic analysis and does not seem to be applicable for large scale epidemiological studies. Moreover, this methodology (due to the intrinsic PCR limitations) is only able to document the predominant population of HBV variants circulating at a given time in a host. Its sensitivity allows detection of genomes which contribute at least with 20-25% of the whole population [Osiowy C., 2006]. Thus, the existence of mixed genotype infections might be underestimated by using PCR amplification followed by direct DNA sequencing and phylogenetic analysis. This important limitation could be overcome only if a considerable number of (PCR amplified) HBV DNA clones had been sequenced for each sample [López et al., 2007].

The phylogenetic analysis is a method to determine the relative and evolutionary relatedness of sequences to each other and control sequences [Mc Cormack et al., 2002]. Its reliability is based on the bootstrapping analysis of 100 to 1000 replicates.

Based on the phylogenetic analysis of (complete or partial) S gene sequences, several research groups have inferred the genotype of HBV isolates from most parts of the world. Based on greater than 4% variation on this gene, most of the recent studies have assigned HBV genotypes in both clinical and molecular epidemiology studies.

The reliability of partial sequences and their phylogenetic analysis largely depend upon the region studied and the lengths of the analyzed amplicons. A potential drawback of this method is the existence of HBV recombinants in other genomic regions, as well as the underestimation of eventual mixed infections.

Other methods currently used for HBV genotype assignment based on limited number of nucleotidic or amino acidic polymorphisms among genotypes, include restriction fragment polymorphism (RFLP) typing, hibridization techniques, genotype-specific PCR amplification, and serological assays.

Once again a potential disadvantage of all these methodologies is the limited genomic information from which a genotype is assigned. It is expected that single nucleotide or amino acid mutations might produce untypeable results. Likewise, an eventual HBV DNA recombination event may not be discovered.

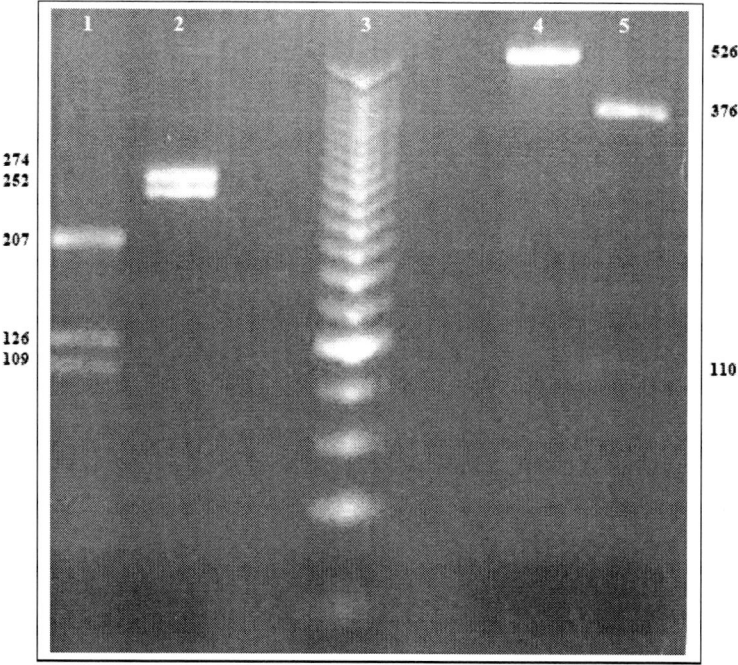

Figure 7. HBV genotyping by RFLP. Two serum samples were analyzed by S gene PCR amplification followed by RFLP [Lindh et al., 1997]. Lanes 1 and 2: PCR products were digested with *Tsp*509I and *Hin*fI, respectively. Restriction pattern ascribes the isolate to genotype A. Lanes 4 and 5: PCR products were cleaved with *Hin*fI and *Tsp*509I, respectively. Digested fragments ascribe the isolate to genotype F. Lane 3: 125 *bp* ladder. The *bp* length of each restriction fragment is shown at both sides of the gel.

Determination of HBV genotype by RFLP. Single or nested PCR amplification followed by endonuclease cleavage is commonly used for RFLP, as described by some groups [Lindh et al., 1997; Mizokami et al., 1999; Zeng et al., 2005]. These methods are able to detect mixed infections, since the endonuclease activities produce cleavage (restriction) fragments characteristic of each HBV populations. However, updates of these methods are required when new genotypes are discovered (figure 7).

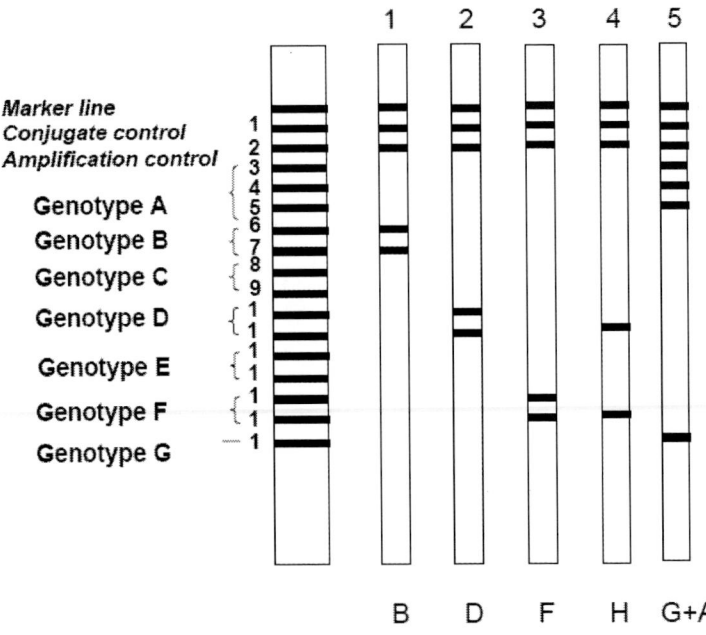

Figure 8. Schematic representation of the Inno-LiPA HBV genotyping assay (Innogenetics). In lines 1 to 5, the genotypes B, D, F, H and G+A mixed infection are determined according to the pattern of reactive bands compared with reference standards.

Determination of HBV genotype by InnoLiPa. A commercially available method was developed by Innogenetics [Teles et al., 1999]. Nested PCR amplification of HBV DNA sequences (by using primers directed to conserved regions of Pre-S1), followed by reverse hybridization of the amplified (biotinylated) DNA products onto genotypic specific oligonucleotides previously fixed on membrane strips, allow proper genotype assignment. The incubation with a streptavidin conjugate allows color development from the biotinylated DNA bound to the strip. Genotypes are determined according to the pattern of reactive bands compared with reference standards (figure 8). This recently available line probe assay (Inno-LiPA HBV genotyping assay, Innogenetics NV, Ghent, Belgium) has been shown to exhibit good correlation with cloning and sequencing data when mixed infections are present in a given sample [Osiowy and Giles, 2003]. As RFLP, this technique can also be applied to study a large number of samples providing representative data for molecular epidemiological studies [Osiowy and Giles, 2003; França et al., 2004].

Determination of HBV genotype by using specific primers for single or multiplex PCR amplification. Naito et al. (2001) have reported the use of a rapid and specific genotyping system for HBV determination based on differences in the conserved nucleotides in the envelope ORFs (Pre-S1 to S). They proposed the use of a universal first amplification (for genotypes A to F, the only known genotypes at the time of their publication) followed by two second round PCR reactions using a mix of primers to differentiate the HBV genotypes based on amplicons' lengths.

Another group has proposed the use of a one-round multiplex PCR to demonstrate the presence of the same genotypes [Kirschberg et al., 2004]. Likewise, Kato et al. (2001) have proposed the use of heminested PCR with specific primers to demonstrate the presence of genotype G, taking advantage of the already known 36-bp insertion in the core gene.

Nevertheless, PCR results should be cautiously analyzed when developed by in-house PCR, since some of the recently published primers may be useless for the detection of some HBV genotypes, as correctly emphasized by Schaefer et al. (2003). These authors alerted the scientific community about possible false negative results for specific genotypes, after analysis of five out of ten recent studies.

Determination of HBV genotype by using serological assays. Several monoclonal antibodies have been obtained in order to characterize HBV genotypes by [Usuda et al., 1999; Usuda et al., 2000; Laperche et al., 2001; Kobashashi et al., 2002; Moriya et al., 2002]. Those monoclonals obtained by Usuda et al. (2000) were raised against the Pre-S2 and S regions, and allow genotyping of strains ascribed to genotype A to F. In a first step, HBsAg present in the serum sample is captured on a 96-well microplate coated with monoclonal antibody directed against the *a* determinant. In a second step, each of four wells receives an enzyme-labeled monoclonal antibody to genotype specific Pre-S2 epitopes (*m*, *k*, *s*, or *u*). Genotypes are determined by the combination of Pre-S2 epitopes. These monoclonals are commercially available.

The co-infection with different genotypes and the recombination between them obviously occur more frequently in geographical regions where a number of genotypes co-circulate and provides the possibility of multiple exposure and superinfection, as well as a mechanism of variation within individuals and in the population in general [Kramvis et al., 2005a, b].

Table 4. Advantages and disadvantages associated to genotyping methodologies

Methodology	Advantage	Disadvantage
PCR amplification, direct HBV full length sequencing and phylogenetic analysis	• *Provides the greatest amount of information: recombinants can be detected.*	• *Minor variants or mixed infections within a given HBV population can be underestimated (sensitivity 20-25%).* • *Not appropriate for large number of samples.* • *Expensive, time consuming, requires significant expertise on phylogenetic analysis.*
PCR amplification, direct HBV partial sequencing and phylogenetic analysis	• *Provides significant information.* • *Useful for analysis of large number of samples.*	• *Recombinants, minor variants and mixed infections can be underestimated.* • *Quite expensive, time consuming, requires significant expertise on phylogenetic analysis.*
PCR amplification + RFLP	• *Useful for massive studies.* • *May detect mixed infections.* • *Easy to perform and relatively cheap.*	• *Requires updates of restriction patterns according to newly discovered genotypes.* • *Recombinants may not be detected*.* • *Polymorphisms at specific positions of endonuclease cleavage may produce untypeable results.* • *Risk of (potential) DNA carryover when (hemi) nested PCR is carried out with large number of samples.*
Hybridization techniques (INNO-LiPA HBV genotyping)	• *Useful for massive studies.* • *May detect mixed infections.* • *Easy to perform.*	• *Quite expensive.* • *Recombinants may not be documented*.* • *Polymorphisms at specific nucleotidic positions may produce untypeable results.*
Genotype-specific PCR	• *Useful for massive studies.* • *May detect mixed infections.* • *Easy to perform and relatively cheap.*	• *Requires updates of designed primers according to newly discovered genotypes.* • *Recombinants cannot be documented*.* • *Polymorphisms at specific positions of DNA template may produce untypeable results, due to lack of complementarity with PCR primers.* • *It might (potentially) produce overestimation of mixed infections if primers used amplify mismatched templates**.* • *Risk of (potential) DNA carryover when (hemi) nested PCR is carried out with large number of samples.*

Table 4. Continued

Methodology	*Advantage*	*Disadvantage*
Genotyping by serological methods	▪ Useful for massive studies ▪ Easy to perform and cheap.	▪ Single amino acidic exchange may turn the analysis impossible or unreliable.

*Unless specifically designed for such purpose.
** Not reported for the currently published protocols for HBV-genotype specific PCR

1.3.2.1. HBV Genotypes and Recombination

As sequence information from full-length HBV genomes is accumulating, recent evidence has undoubtedly confirmed that this is not an infrequent event. HBV genotypes coinfections have been documented (i.e. A + G or A+ D). Moreover, a given predominant genotype in a host may be replaced by a minor one during the course of the infection [Chen et al., 2006]. If a given cell is coinfected by two (or more) different genotypes, the possibility exists that a recombinant genome is being synthesized. With our current knowledge of HBV replication, no mechanism can be envisioned where two hepadnaviral genomes can exchange their genetic material at the level of transcription [Schaefer S., 2007]. Initial evidence for recombination was independently provided by two groups in the early nineties, when studying consecutive samples from a chronic infected patient [Tran et al., 1991] and those from patients with hepatocellular carcinoma [Georgi-Geisberger et al., 1992]. Subsequently, incongruities when studying partial sequences derived from different genomic regions [Norder et al., 1994] made scientists suspect that these might be true recombinants. This biological event was formally proven after Bollyky et al. (1996) identified specific breakpoints in published sequences. Since then, recombination events have involved genotypes such as A/C, A/D, A/E, B/C, B/D, C/D, C/F, C/G and C/U (U for unknown genotype) [Bowyer and Sim, 2000; Morozov et al., 2000; Hannoun et al., 2000; Yuasa et al., 2000; Cui et al., 2002; Owiredu et al., 2002; Sugauchi et al., 2002b; Fares and Holmes, 2002; Kurbanov et al., 2005; Wang et al., 2005; Simmonds and Midgley, 2005; Suwannakarn et al., 2005; Chen et al., 2006; Olinger et al., 2006]. An interesting example of recombination was documented between genotypes B and C, which led to the discovery of the so-called Ba subtype (currently B2), geographically restricted to the Asian continent without affecting Japan. In contrast, a genuine (pure) genotype B genome circulates in Japan and was formerly classified as genotype Bj (currently B1) [Sugauchi et al., 2002a]. As observed when comparing B1 and B2 subtypes, A1/A2 and C1/C2 pairs differ significantly in many virological and probably some clinical associated features [Schaefer S., 2007].

Further recombination events between HBV genotypes have very recently been reported: i.e. A/B/C, A/G and B/C/U.

Genotypes A and C showed the highest recombination tendency. Region priority and breakpoint hot spots in the intergenotype recombination have been demonstrated. Recombination breakpoints were found to be concentrated mainly in the vicinity of the DR1 region (nt 1640-1900), the pre S1/S2 region (nt 3150-100), the 3'-end of the C-gene (nt 2330-2450) and the 3'-end of the S-gene (nt 650-830). Again, in this study it was suggested that intergenotype recombinants may result from co-infection with different genotypes [Yang et al., 2006]. A general comprehensive view of HBV genotypes and their recombinant forms has recently been published [Schaefer S., 2007]. At least one more putative genotype would remain to be discovered [Yang et al., 2006; Schaefer S., 2007].

1.3.2.2. HBV Genotype and Its Clinical Relevance

It has become more evident that the long-term prognosis, the initial clinical picture and the response to treatment may differ depending on which genotype/s have infected the patient [Kidd-Ljunggren et al., 2004; Kramvis et al., 2005a]. Although additional studies are needed to better define the role of genotypes in viral replication, carcinogenesis, fibrosis progression, and antiviral response, HBV genotyping should be considered as a necessary procedure before initiating any treatment.

However, it is increasingly difficult to extrapolate findings from one geographical area to another, as the course of HBV infection also depends on both host and environmental factors. Therefore, larger studies are necessary in various regions of the world. Most of the available information comes from studies carried out in Southeast Asia, where HBV infection is hyperendemic and genotypes B and C predominate. These analyses have recently been extended to other regions, but data from Latin America is still scarce.

Patients infected with genotype B are more likely to have a sustained biochemical remission after spontaneous HBeAg seroconversion than patients infected with genotype C [Chu et al., 2002], who are more prone to develop chronic and advanced liver disease [Sugauchi et al., 2002; Ishikawa et al., 2002]. Associations were described between genotype C infection and higher scores of histological activity [Lindh et al., 1999; Lindh et al., 2000], higher ALT levels and hepatocellular carcinoma than in cases of infection with genotypes B [Sugauchi et al., 2002; Ding et al., 2001], D or A [Vivekanandan et al., 2004].

Chronic hepatitis appeared to be more frequent in patients infected with genotype A than in those infected with genotype D [Mayerat et al., 1999]. However, chronically infected patients with genotype A may have a better prognosis than those infected with either genotypes D or F due to spontaneous sustained biochemical remission and milder disease with a decrease of HBV-DNA levels at a higher rate [Sanchez-Tapias et al., 2002].

Genotype D was also found to be more related to a severe recurrent disease post-transplantation [McMillan et al., 1996]. In a single study, the analysis of a small group of Spanish patients infected with the Amerindian genotype F suggested that this genotype may be associated with a higher mortality rate because of a more frequent liver failure [Sánchez-Tapias et al., 2002].

Moreover, a very recent study demonstrated that genotype F was more prevalent among patients with hepatocellular carcinoma (HCC) (68% versus 18%) than among those without HCC ($p<0.001$). Even more importantly, for patients infected with genotype F, the median age at diagnosis of HCC was lower than that for patients with other genotypes (22.5 vs. 60 years, respectively; $p=0.002$) [Livingston et al., 2007]. These results clearly emphasize the genotype relevance in the outcome of HBV infection.

Another interesting point is the age of HBeAg to anti-HBe seroconversion among the different HBV genotypes. It has been reported that with genotypes A and D, the age of seroconversion to anti-HBe is intermediary and lies between a seroconversion at very early age with genotype E and at very late age with genotypes B and C [Allain JP., 2006].

Very few isolates of HBV genotype G or H have been characterized, making it difficult to draw any conclusion regarding the influence of any of them on disease progression or response to antiviral treatment. Chronic hepatitis patients infected with genotype G are characterized by high HBV-DNA, HBeAg [Vieth et al., 2002; Kato et al., 2002b] and ALT levels [Kato et al., 2002a]. However, the frequent co-infection with genotype A might be responsible for these attributes [Kato et al., 2002a; Kato et al., 2002b].

Data on response to antiviral treatment among different HBV genotypes are preliminary. It seems that a higher rate of HBeAg seroconversion follows after interferon (IFN)-α treatment in patients infected with genotype A than in those with genotype D (37% vs. 6%) [Erhardt et al., 2000]. Another study showed that, in addition to low pre-treatment HBV-DNA and elevated ALT levels, genotype B was associated with a better antiviral response to IFN-α treatment. A higher rate of HBeAg seroconversion was recorded when compared with genotype C [Wai et al., 2002].

The effect of HBV genotype on response to nucleoside and nucleotide analogs is unclear due to the paucity of data. HBV genotype does not appear to correlate with adefovir or lamivudine treatment response, but again, the studies regarding treatment with both of these drugs are few and preliminary.

One study showed that although genotypes B and C have a similar risk of developing lamivudine resistance after 1 year of therapy, genotype B has a better virological response than genotype C [Kao et al., 2002].

A very interesting German study showed that the risk for the emergence of lamivudine resistance during the first year of treatment is 20 times higher in HBV infections with subtype *adw* than with *ayw*. A possible explanation for the reduced risk of developing resistance in subtype *ayw* could be associated directly or indirectly to a change in the hydrophilicity within the S region of this subtype but not in *adw* [Zollner et al., 2001]. However, when patients were compared after 2 or 3 years of lamivudine treatment, no difference was observed between patients infected with subtype *adw* or *ayw*. As a conclusion, lamivudine resistance mutations took longer to develop in subtype *ayw* [Zollner et al., 2002]. The same group documented that HBV genotype (D or A) and pretreatment serum HBV DNA were independently associated with emerging rt204I or rt204V (formerly Met552Ile or Met 552Val) lamivudine resistant mutants [Zollner et al., 2004].

On the other hand, there was no significant difference in the antiviral response to adefovir among patients infected with the different HBV genotypes [Westland et al., 2003].

1.3.2.3. HBV Genotypes (and Their Variants and Mutants): Influence on Diagnosis

The analytical sensitivity of HBsAg and anti-HBs assays may be dependent on HBV genotype or subtype. Preliminary results show that new real-time Nucleic Acid Testing (NAT) detects genotypes A to G with equal sensitivity [Weber B., 2005]. More recently, an A-H genotype independent real time PCR was developed [Liu et al., 2006]. The clinical significance of S-gene mutants and other HBV mutants needs -in analogy to that of HBV genotypes- to be further investigated. Although many mutant strains of HBV have been identified, the most clinically significant strains include the Pre-C and core-promoter variants, as well as the surface and polymerase mutants. Interpretation of the clinical significance of HBV mutations is complicated by the lack of standardized nomenclature, variable sensitivity in the assays used to detect them, and presence of multiple mutations even in the same species. Despite these limitations, *in vivo* analysis of naturally occurring HBV variants and *in vitro* mutagenesis studies have identified several variants that may play a role in disease pathogenesis, immune escape, and resistance to antiviral therapy [Wai et al., 2004].

Very few studies have addressed the issue of the potential influence of HBV genotypes on the sensitivity of molecular assays. In one of them, in order to assess the sensitivity of diagnostic assays, a panel (from the Société Française de Transfusion Sanguine, Nord-Pas-de-Calais, France) was used for testing dilution series of HBV subtypes. Approximately, a 10-fold difference in sensitivity was recorded with three commercial assays [Weber et al., 2003; Weber B., 2005]. A subsequent collaborative study from this author was carried out with 200 serum samples from Nigerian liver patients by using three different HBsAg assays and HBV DNA real time PCR. Again, one assay showed a higher sensitivity than the alternative tests. By sequencing the DNA from one of the positive samples detected with the former assay, a genotype E isolate was demonstrated [Weber et al., 2006]. These initial and preliminary results suggest that there is a genotype dependence on the sensitivity of the HBsAg assays, especially with genotype E, the most divergent genotype at the *a* determinant. In this regard, a recent study showed that 9 out of 10 commercially available kits for HBsAg detection, including chemiluminescent immunoassays (CLIA), enzyme immunoassays (EIA), and chemiluminescent enzyme immunoassays (CLEIA) were able to demonstrate 0.2 IU/ml of such antigen, from all genotypes. However, the remaining one (an EIA using a polyclonal for capture and a monoclonal for detection) failed to detect such amount of HBsAg from genotype E and F [Mizuochi et al., 2006]. Nevertheless, the sensitivity recorded with all kits (0.2–1.0 IU/ml) approaches the satisfactory criterion according to the "Guidance for Industry" issued by the FDA or the Common Technical Specification ("CTS") defined by the European Union. Clearly, a more in-depth analysis with the most divergent genotypes needs to be performed regarding this important issue.

With regard to S gene variants/mutants, they can be didactically classified into three groups: a) immune escape mutants; b) diagnostic escape mutants; and c) immune and diagnostic escape mutants. These topics will be discussed below (section 3.1).

Chapter 2

HBV Epidemiology in Latin America and the Caribbean (LAC) Region

2.1. Current Theories about HBV Origin and Its Distribution

One of the theories that account for the geographical distribution of HBV genotypes proposes that HBV originated in America, and spread into the Old World during the European colonization over the last 400 years. It was suggested that the existence of the various human HBV genotypes were the result of the diversification of HBV after their geographical dispersal over such 400-year period, although the date calculated from the molecular clock implies that a much longer period would be required (over 2000 years). In spite of the probability that the human HBV genotypes pre-existed before the spread, there is no evidence of infection other than with genotype F in contemporary indigenous populations in South America.

However, the fact that HBV is widely distributed in Old World primate species, such as chimpanzees, gorillas, orangutans and gibbons, makes it difficult to accept the previous hypothesis. Furthermore, although primate infection was initially thought to have being the result of accidental transmission from humans to captive animals, it has now been firmly established that chimpanzees from West Africa and gibbons and orangutans from Southeastern Asia are infected with HBV in the wild. Taking into account the predictions of the molecular clock, the primate genotypes of HBV would also have originated within the last 2000-3000 years. The main difficulty to provide an alternative hypothesis for the evolutionary history of HBV is the unexpected fact of an equivalent sequence relationship among human genotypes A-E and G to each other and to the primate-species-associated genotypes of HBV. It is also difficult to rationally fit into any scheme the divergent human HBV genotype F and the even more divergent HBV variant obtained from a captive woolly monkey, a New World primate. The existence of species-specific genotypes of HBV in chimpanzees, gibbons and orangutans, and the way they are intermixed with human genotypes is unexpected, as the primate viruses should be much more divergent from human variants and from each other, given the much longer period of co-speciation of primate species (10-15 million years).

The third hypothesis for HBV origins proposes that human HBV infection possibly arose mostly through contact with different primates infected with species-specific genotypes (equivalent to those found in chimpanzees, gibbons and orangutans). In some way, this is quite similar to what it is believed for the origins of HIV infection in humans. Infection with HIV-1 is likely to have originated through at least three separate cross-species transmissions from chimpanzees, while human infection with HIV-2 in Western Africa independently arose several times through contact with sooty mangabeys. A primate origin for human HBV infection is indeed supported by the observation that the areas of high HBV prevalence in humans are those in which contact and cross-species transmission from primates is most likely (South America, sub-Saharan Africa and southeastern Asia). In fact, certain HBV genotypes are mainly restricted to these three areas (F, E and B/C, respectively). Importantly, (genotype C) human and ape HBV recombinants have been documented to circulate among humans in Africa and Asia [Simmonds and Midgley, 2005].

2.2. Epidemiological Patterns and Seroprevalence Rates Vary among Geographical Regions and Socioeconomic Groups

Over 400 million people live in the LAC region, among whom more than 6 million are chronic carriers of HBV and 400,000 new infections occur every year [Fay et al., 1990; Torres JR., 1996]. Therefore, HBV represents a major public health problem and is a cause of significant morbidity and mortality, especially

HBV Epidemiology in Latin America and the Caribbean (LAC) Region   33

for the communities living in the Amazon basin, as more than 30% of the 4 million carriers in South America is located in this region [Fay et al., 1990].

Figure 9. Chronic hepatitis B distribution in Latin America and the Caribbean region.

The worldwide distribution of chronic HBV infection shows 3 different patterns (table 1) that can also be distinguished among LAC countries [Tanaka J., 2000] (figure 9):

- *High Endemicity Areas*: Amazon Basin (parts of Northern Brazil, Colombia, Peru and Venezuela).
- *Intermediate Endemicity Areas*: Haiti and Dominican Republic in the Caribbean region; Guatemala and Honduras in Central America; and the Northern part of Brazil, Colombia, Surinam and Venezuela in South America.

- Low Endemicity Areas: Mexico in North America; Bahamas, Barbados, Cuba, Grenada, Jamaica and Trinidad and Tobago in the Caribbean region; Costa Rica, El Salvador, Nicaragua and Panama in Central America; and Argentina, Bolivia, the Southern part of Brazil, Chile, Ecuador, Paraguay and Uruguay in South America.

Data on the prevalence of HBV infection in different regions of LAC are sparse and originate from numerous local and regional serological studies. Moreover, interpretation of these data is also complicated due to the different methodologies used to asses HBV infection among studies and the fact that the non-uniform pattern of distribution of HBV infection is related to geographical, social and cultural factors.

The emigration of people into the LAC region and the migration of populations throughout each country can also give rise to rapid changes in epidemiological patterns. For example, the city of Buenos Aires (Argentina) as well as the southern and southeastern regions of Brazil have experienced increasing rates of immigration from high endemicity areas. Continued urbanization and increasing access to small isolated municipalities have contributed to rising HBV endemicity throughout LAC countries [Gish et al., 2006].

Unless otherwise specified, data regarding HBV seroprevalence mentioned in the forthcoming Sections correspond to blood donors. These values do not necessarily reflect those from the general population (i.e. underestimations related to the age of donors). Nevertheless, many studies have assumed the observed values among blood donors as representative of the general population.

2.2.1. HBV Epidemiology in Mexico

The prevalence of HBsAg and anti-HBc positive blood donors from several Mexican states ranges from 0.1 to 0.3% and from 1.4 to 21.9%, respectively [Alvarez-Muñoz et al., 1991]. These rates usually increase after the age of 16, suggesting the presence of a horizontal transmission [Tanaka J., 2000].

Interestingly, the prevalence of HBV infection in Mexican prison inmates and female sexual workers were comparable to those of the general population and blood donors [Alvarado-Esquivel et al., 2005 and 2006; Juárez-Figueroa et al., 2006].

However, Guanajuato, with 1.1% of its blood donors showing positivity for HBsAg, is one of the Mexican states with the highest prevalence rates of HBV infection [Carreto-Velez et al., 2002]. Another interesting example is the southern state of Chiapas, where Guatemalan refugee camps are spread among the local communities and 4.2 and 17.3% of HBsAg (+) was found among Mexicans and Guatemalans, respectively. Besides, HDV was present in 42.0% of the HBsAg carriers. The HBV prevalence rates in southeastern Chiapas could easily be compared to those in highly endemic areas, like the Amazon Basin [Alvarez-Muñoz et al., 1989].

In more recent studies, occult HBV infection was reported with a prevalence of 8.2 and 18.4% among Mexican blood donors and HIV (+) patients negative for HBsAg, respectively [Garcia-Montalvo et al., 2005; Torres-Baranda et al., 2006].

2.2.2. HBV Epidemiology in the Caribbean Region

In the period 1996-2003, the prevalence rates of HBsAg among blood donors from several Caribbean countries were studied, showing figures that ranged from 3.0 to 7.5% in Saint Kitts and Nevis; between 4.1 and 5.5% in Haiti; 1.6 to 2.5% in Grenada and Guyana; remained consistently above 1.0 in the Bahamas, Saint Vincent and the Grenadines; and below 1.0% during the entire period of time in Cuba, Trinidad and Tobago, Antigua and Barbuda and Suriname [Cruz et al., 2005].

In the Dominican Republic, the overall seroprevalence rate of anti-HBc among 478 residents was 21.4%, with a high seroprevalence rate in children and even higher in those aged over 16 years old. Interestingly, this rate was higher in females than in males (24.0% vs. 12.6%, $p = 0.01$) [Tanaka J., 2000].

Among Haitian women known to be infected with human immunodeficiency virus type 1 (HIV-1) or Human T Lymphotropic Virus type I (HTLV-1), the seroprevalence rates of anti-HBc were 67.0 and 43.0%, respectively [Boulos et al., 1992].

In Puerto Rico, the prevalence for HBsAg and anti-HBc in the 1980's was 10.1 and 0.2% [Mazzur et al., 1981]. On the other hand, the prevalence of HBV serological markers is higher among Cuban high-risk groups, such as multi-transfused [Ballester et al., 2005] and HIV [+] patients [Rodriguez et al., 2000] as compared to the general population. Only 1.3% of high-performance athletes showed positivity for the HBV infection [Rodrigues Lay et al., 1997] while the seroprevalence rate of anti-HBc among the risk groups was 45.5%, being 5.1% of them positive for HBsAg. Significant statistical association was found between the male sex and the homosexual or bisexual behaviour to the exposure to HBV (p <0.01 and p < 0.001, respectively).

2.2.3. HBV Epidemiology in Central America

The overall seroprevalence rates for HBV infection in Costa Rica ranged from 0.4 to 0.5% blood donors thus showing a low prevalence (period 1993-1997) [Schmunis et al., 1998; Schmunis et al., 2001]. Groups other than blood donors were studied as well: the mean HBsAg prevalence was 5.0, 18.9, 43.2 and 54.3% among cirrhotics, haemophiliacs, patients with viral chronic active hepatitis and hepatocellular carcinoma, respectively [Salom et al., 1990].

In El Salvador, the overall seroprevalence rates for HBsAg ranged from 0.4 to 0.8% in blood donors (period 1993-1997), while in Guatemala it reached 0.7% (1993) and in Honduras it oscilated between 0.3 to 1% (period 1993-1997) [Schmunis et al., 1998; Schmunis et al., 2001]. Moreover, in the latter country, multi-transfused patients from the cities of Tegucigalpa and San Pedro Sula were analyzed between September 2002 and August 2003, and 11% were found positive for HBsAg and 27% for anti-HBc [Vinelli et al., 2005].

The prevalence rates for HBV infection in Nicaragua varies between 0.3 and 0.8% in blood donors (period 1993-1997) [Schmunis et al., 1998; Schmunis et al., 2001]. In 1996, the seroprevalence for markers of HBV infection was studied among a healthy population in Leon. Only 6% and 0.9% showed anti-HBc and HBsAg (+), respectively [Perez et al., 1996]. These findings contributed to confirm that Nicaragua is a low endemicity country.

The overall seroprevalence rates for HBV infection in Panama ranges from 0.4 to 0.7% (period 1994-1997) [Schmunis et al., 2001]. On the other hand, in Belize, a small country in Central America, it is 1.0 to 4.7% in blood donors (period 1996-2003) [Cruz et al., 2005]. Moreover, 31% of the members of the National Defense Force were positive for anti-HBc and 4% had HBV surface antigen.

Rates of anti-HBc increased with age and varied significantly among the ethnic groups with the lowest rates among Mestizo (5%) and Mayan Indians (9%), and the highest in Creoles (30%) and Garifuna ones (56%) [Craig et al., 1993]. Similar seroprevalence rates were recorded when health care workers were studied [Hakre et al., 1995].

2.2.4. HBV Epidemiology in Colombia

In 1985, the seroprevalence of HBV infection among healthy blood donors from the urban area of Bogota was 7% for anti-HBc and 1.6% showed positivity for HBsAg [Echevarria et al., 2003]. Moreover, it ranged between 0.7- 0.9 % during the period 1994-1997 [Schmunis et al., 2001]. However, these figures are lower compared to those observed among communities in northwestern Colombia belonging to the Amazon Basin.

"Hepatitis of the Sierra Nevada de Santa Marta" has been recognized as an unusual type of severe hepatitis occurring in this particular area of Colombia since 1930. This endemic form of acute hepatitis occurred predominantly in several small towns within 50 km of Santa Marta, most commonly affecting children under the age of 15 and males twice as frequent as females (figure 10). This distinct type of fulminant hepatitis became a serious health problem in the Colombian Amazon due to its high mortality rate, which in the 1940's reached 1.25 per 1,000 inhabitants per year [Buitrago et al., 1986].

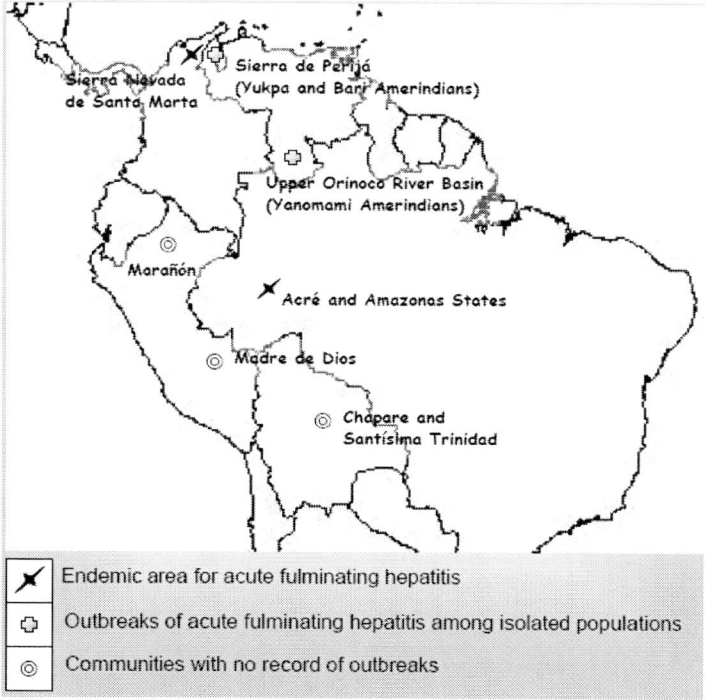

Figure 10. HBV/HDV coinfection areas in Latin America.

The anti-HBc prevalence in this area in 1985 was up to 93%, with an HBsAg incidence of over 20%. In one particular community of the three analyzed, 60% of the HBsAg carriers were positive for anti-HDV [Ljunggren et al., 1985].

Several studies from the 1980's [Hadler et al., 1984; Ljunggren et al., 1985; Bensabath et al., 1987; Torres et al., 1991] demonstrated an association between the occurrence of endemic cases or epidemic outbreaks and the existence (in the same region) of HDV superinfection among chronic HBV carriers in highly endemic areas for the latter. Although the etiology of Santa Marta hepatitis is now known, no clear conclusions can be drawn about the viral transmission mechanisms driving this deadly form of hepatitis.

2.2.5. HBV Epidemiology in Venezuela

The overall seroprevalence rate of anti-HBc in Venezuela is 3%, but increasing after the age of 16 [Tanaka J., 2000]. However, variations are frequent due to geographical and cultural factors. While endemicity is considered to be medium among rural populations and low in urban areas, it is extremely high in Amerindian communities from the Amazon State and Sierra de Perija (Zulia State).

In the metropolitan area, the prevalence of HBsAg among healthy subjects [de Marquez et al., 1993] and indigent communities [Ponce et al., 1994] was not significantly different (1.6% vs. 2.1%). Anti-HBc was higher among indigent patients (28.0% vs. 10.6%). These differences appear to correlate with the socio-economic level. Interestingly, similar conclusions were drawn when HBV infection was compared in pregnant women residing in Caracas and belonging either to a low or a high income population [Pujol et al., 1994].

Rural populations have a prevalence rate for HBsAg of 2.2%, which is significantly higher than that observed in urban groups but no different from the one in Afro-Venezuelan communities (3.6%) [Quintero et al., 2002].

By the late 1960's, an outbreak of acute fulminating hepatitis occurred among the Yanomami, an isolated Amerindian population of the Upper Orinoco river basin (southern Venezuela), with HBsAg and HDV positive in 30.6 and 39.7% of the studied patients. The epidemics extended for several years and reached severe proportions in 1975 with 320 cases per thousand inhabitants and several deaths [Torres et al., 1991].

A similar outbreak affected the Yukpa Indians in the Perija region (northwestern Venezuela, figure 10) a decade later, where 65% of patients tested positive for HBsAg and 86% of them had HDV coinfection. This new epidemic extended from 1979 to 1981, primarily affecting children and young males and causing 149 acute cases, 22 of which developed chronic hepatitis, and 34 ended in death [Hadler et al., 1984]. Moreover, when the Yukpa HBV carriers who survived the 1979-1981outbreak were followed regularly between 1983 and 1988, half of them tested positive for HDV infection. This particular group was at a higher risk of developing moderate-to-severe chronic liver disease with mortality rates ranging from 6.9 to 8.8% per year, compared to HBV carriers without HDV coinfection [Hadler et al., 1992]. Interestingly, these high endemicity rates for HBV and HDV infection were not exclusive of the Yukpa Indians as a similar epidemiological description was reported among the Bari Indians also living in the Perija mountains area [Blitz-Dorfman et al., 1994]

In both cases, these Amerindian communities located so far away one from the other shared the same etiology: HDV superinfection among chronic HBV carriers from highly-endemic areas (figure 10).

In more recent studies, HBV DNA was detected in blood samples positive for anti-HBc and negative for HBsAg. The prevalence of "occult" HBV infection ranged between 1.4% to 6.0% among blood donors in Caracas [Gutiérrez et al., 1999; León et al., 1999; Gutiérrez et al., 2004]. However, no HBV DNA was found in the sera of blood donors who were negative for all HBV serological markers [Gutiérrez et al., 2001].

2.2.6. HBV Epidemiology in Ecuador

The seroprevalence rates for HBV among Ecuadorian blood donors ranged between 0.4 and 0.5% during the period 1994-1997 [Schmunis et al., 2001].

2.2.7. HBV Epidemiology in Brazil

The overall seroprevalence for anti-HBc was 7.9%, showing an increase after the age of 30 years and was higher in males. Remarkably, 3.8% of the children aged between 1 and 5 years old were positive for anti-HBc, which suggests the occurrence of some degree of vertical transmission [Tanaka J., 2000].

Brazil is a country with an overall intermediate HBV prevalence, which shows increasing rates from south to north [de Paula et al., 2001]. The pattern of distribution of HBV infection tends to show a wide variation. It depends on the geographical area where the study is carried out as well as on the socio-economic level of the population group analyzed. In general, the seroprevalence is higher among the lower income groups and the communities living in northern Brazil, especially in the Amazon Basin [Tanaka J., 2000].

In the southern states (Paraná, Rio Grande do Sul, Santa Catarina), the seroprevalence of HBsAg and anti-HBc among blood donors samples collected during the period 1999-2001 ranged from 0.6 to 1% and 5.3 to 8.8%, respectively [Rosini et al., 2003]. As expected, the rates were higher among haemodialysis patients (10.0% for HBsAg and 23.2% for anti-HBc), health care workers (2.7 and 20.1%) [Carrilho et al., 2004] and HIV infected patients (24.3 and 71.2%) [Treitinger et al., 1999].

During the 1998-2005 period 0.3% and 3.7% of the blood donors in Rio de Janeiro (South-East of Brazil) showed positivity for HBsAg and anti-HBc, respectively [Andrade et al., 2006]. Interestingly, the rates among the indigenous population of this region (Xacriabá) are comparable to those from blood donors residing in urban areas [Figueiredo et al., 2000]. On the other hand, the prevalence of HBsAg was 4.4% among hemodialysis patients [Busek et al., 2002], 3.4% in injecting drug users [Oliveira et al., 2005] and between 5.3 and 8.5% in HIV infected patients [Mendez-Correa et al., 2000; Pavan et al., 2003; Souza et al., 2004)].

The central-west Brazil shows higher HBV prevalence in blood donors (9.4% for anti-HBc and 0.7% for HBsAg) [Aguiar et al., 2001], than among isolated Afro-Brazilian (2.4 and 2.2%, respectively) [Motta-Castro et al., 2005] and indigenous communities (2.2 and 0.0%) [Aguiar et al., 2002]. However, as stated previously, HBV markers are easily detected among individuals from high-risk groups, such as hemodialysis patients (29.8% for anti-HBc and 0.8% for HBsAg) [Ferreira et al., 2006] and HIV infected patients (40.0 and 3.7%) [de Almeida Pereira et al., 2006].

The northern region includes the Amazon River basin, which has one of the highest rates of HBV carriers in the world (5-20%) and a mortality rate which is 5 to 10 times higher than the averages for the rest of the hemisphere [Bensabath et al., 1987; Soares et al., 1994; Dutra Souto et al., 1998]. In fact, a high frequency of severe cases of acute and chronic hepatitis is commonly observed and even an endemic and unique form of fulminant HBV/HDV hepatitis, known as "Labrea" fever [Fay et al., 1985; Bensabath et al., 1987] has been reported. This type of fulminant hepatitis is characterized by liver failure before hepatocellular necrosis and inflammation are observed and resembles another peculiar form of fulminant hepatitis caused by HBV/HDV superinfection in the African equatorial forest (Bangui hepatitis) [Andrade et al., 1992] (figure 10).

In western Brazilian Amazon (Acre and Amazonas states), the overall HBsAg prevalence among blood donors was 66.1%. This high endemicity is complicated by the high prevalence of HDV (66%) [de Paula et al., 2001]. On the other hand, in eastern Amazon (Pará state), the picture is slightly different. HBV infection began in the first two years of life, resulting in up to 14.4% of carriers and 85.0% of the indigenous population with signs of past infection. However, no HDV co-infection has been detected yet [Soares et al., 1994].

In recent studies, "occult" HBV infection has been observed among different Brazilian groups. HBV DNA was detected in 3.3% of blood donors [Silva et al., 2005], 1.0% of renal transplant patients [Peres et al., 2005], 14.0% of chronic HCV patients [Silva et al., 2004] and 18.7% of HIV infected patients [Sucupira et al., 2006].

2.2.8. HBV Epidemiology in Peru

The overall HBsAg seroprevalence rates for HBV infection among Peruvian blood donors ranged from 0.7 to 1.0% during the period 1993-1997 [Schmunis et al., 1998: Schmunis et al., 2001].

The epidemiology of the HBV infection in this Southern American country is very rich due to regional variations in the viral prevalence:

- The Peruvian Amazon plain (Marañon and Madre de Dios; figure 10) is a highly endemic region where HBV infection occurs mainly among individuals younger than 20 years old. The prevalence of anti-HBc among the autochthonous population ranges from 69.0 to 74.0%. Moreover, between 3.9 to 12.1% are positive for HBsAg, among whom 2.5-9.0% has anti-HDV [Echevarria et al., 2003]. Similar to what happened in other regions of the Amazon Basin, the HBV/HDV coinfection (highly endemic along the Northern and Northeastern border of the Peruvian Amazon jungle) was found associated to outbreaks of severe acute hepatitis occurred during 1992-1993 among troops stationed at four jungle outposts [Casey et al., 1996] (figure 10).
- The Andean highland is a medium to low endemic area. For example, 9.4 and 1.4% of pregnant women were found positive for HBsAg in the cities of Lima and Chanchamayo, respectively [Vasquez et al., 1999]. This conflicting result was explained due to the fact that 63% of pregnant women attending the Lima Hospital were derived from regions other than the capital city.
- The transition valleys (Huanta and Abancay), communicating the previous regions, are medium to high endemic areas. Therefore, it was

not surprising that when 143 clinically healthy school children were studied, 82.0%, 16.0% and 17.9% showed positivity for anti-HBc, HBsAg and anti-HDV, respectively. However, so far, no outbreaks have been reported in the area [Cabezas et al., 1994].

2.2.9. HBV Epidemiology in Bolivia

In general, Bolivia is considered to be a medium to low HBV endemicity country, with an overall HBsAg prevalence ranging from 0.4 to 8.0%. However, it depends on the population group and the geographical region studied.

HBV infection is lowly-intermediately endemic among healthy blood donors residents of Santa Cruz city or its suburbs [Konomi et al., 1999], rural communities of the high Andean plateau, and two high-risk groups in the city of Cochabamba (homeless children and sexual workers) [León et al., 1999] as 0.3, 11.2 and 11.6% of the analyzed population was found positive, respectively.

On the other hand, a very different picture is observed when the indigenous populations of the Bolivian Amazon (Yuki Indians in Chapare and Trinitarios Indians in Santisima Trinidad; figure 10) become the subject of study. As has been noted previously with similar communities in tropical areas of South America, HBV infection is highly endemic with an anti-HBc prevalence of 34.0-84.0% mainly among individuals younger than 20 years old. Up to 4.8% of the population are HBsAg carriers, among whom the prevalence of anti-HDV was significantly low (0.0-2.2%). So far, no outbreak of HDV infection has been documented in this region, but the high endemicity of HBV points to the possibility of future outbreaks [León et al., 1999; Echevarria et al., 2003].

2.2.10. HBV Epidemiology in Paraguay

The seroprevalence rates for HBsAg among blood donors ranged from 0.6 to 1.4% during the period 1994-1997 [Schmunis et al., 2001].

2.2.11. HBV Epidemiology in Uruguay

The seroprevalence rates for HBsAg among blood donors ranged between 0.4 and 0.5% during the period 1994-1997 [Schmunis et al., 2001]. A survey performed on injecting drug users (IDUs) and non-injecting cocain users (NICUs) between 2002 and 2003 established an anti-HBc seroprevalence of 19.5 and 11.7%, respectively (Vignoles et al., 2006).

2.2.12. HBV Epidemiology in Chile

HBsAg prevalence is low (about 0.1%) among blood donors in Chile [Schmunis et al., 2001] with variations within the country: the northern area shows higher rates than central and southern Chile [Velasco et al., 1972].

Santiago, the Chilean capital city, exhibits a low endemicity rate also. Two studies support this fact. When 117705 voluntary blood donors from the National Blood Bank were studied, only 0.3% of them were chronic HBsAg carriers [Velasco et al., 1978]. More recently, the overall seroprevalence rate of anti-HBc was 0.6% among the 496 Chileans analyzed with ages ranging from 1 to 40 years. An increase in this rate was observed after the age of 20, with the highest prevalence recorded in the group ranging from 31 to 40 years (4%) [Tanaka J., 2000]. However, taking into account these features, it is plausible that a higher anti-HBc prevalence rate could have been observed if an older group had been included as well.

2.2.13. HBV Epidemiology in Argentina

During the year 2005, the overall seroprevalence for HBsAg and anti-HBc among blood donors was 0.3 and 3.1%, respectively [Proyecto Programa Nacional de Control de Hepatitis Virales, Informe Epidemiológico N° 6, 2006]. The prevalence rates for anti-HBc tend to increase after the age of 30 years, but are significatively present (3.8%) among children aged up to 5 years, whereas it decreases when older children and young teenagers are studied [Tanaka J., 2000]. This finding suggest the occurrence of some degree of vertical transmission in Argentina, which was later indirectly confirmed by the presence of detectable HBsAg among 0.2% of pregnant women in 2004 and 0.1% in 2005 [Proyecto Programa Nacional de Control de Hepatitis Virales, Informe Epidemiológico N° 5 (2005) and N°6 (2006), respectively]. Nevertheless, these values may underestimate the true HBsAg prevalence among pregnant women, since they were strongly influenced by the high number of pregnancies studied in one hospital located in a region (Córdoba Province, central area) of very low prevalence (0.1%). Other hospitals have shown HBsAg prevalence rates reaching 0.5% (Chaco Province, northern area).

As expected, rates were significantly higher among high-risk groups, such as HIV infected population (58.5% and 14.5% positive for anti-HBc and HBsAg, respectively [Fainboim et al., 1999]). Likewise, the anti-HBc seroprevalence rates among men having sex with men -MSM- (37.7%), female sex workers (14.4-17.9%), individuals with other sexual infections (15.0%), NICUs (8.9%) and street-recruited injection drug users (42.5%) were also high [Weissenbacher et al., 2003; Pando et al., 2006a and 2006b; Vignoles et al., 2006]. Occult HBV infection was detected among 2.0% of HBsAg (-) and anti-HBc (+) patients coinfected with HIV and/or HCV [Munne et al., 2006].

Although the few epidemiological studies available place Argentina as a low endemicity country, regional variations are common. The northern area of this country shows a higher prevalence (3.6% of the blood donors in the east and 7.2% in the west are positive for anti-HBc) when compared to the central and southern regions [Proyecto Programa Nacional de Control de Hepatitis Virales, Informe Epidemiológico N° 6, 2006]. Two interesting examples are represented by the north-western provinces of Salta and Jujuy. Salta and San Salvador de Jujuy cities are considered to be low to intermediate HBV endemicity areas taking into consideration that the HBsAg and anti-HBc prevalence is 1.0 and 9.3%, and 1.2 and 8.4%, respectively. However, Orán and Embarcación (near the Argentinean-Bolivian frontier) in Salta (4.1 and 41.8%, and 3.5 and 33.8%, respectively) and the Yungas tropical area in Jujuy (2.6 and 17.4%) show the highest HBV seroprevalence not only of these provinces, but also of the entire country [Garay et al., 2006; Remondegui et al., 2006].

2.3. HBV Molecular Epidemiology in the LAC Region: A Unique Genetic Diversity as a Consequence of a Continuous Human Migration

It has been proposed that indigenous HBV genotypes entered the American continent with the first settlers that came from Asia across the Bering Strait around 20,000 and 15,000 years ago [Arauz-Ruiz et al., 2002]. They probably migrated to the south along the coastal area, since the global coalescence prevented access to North America through any interior route. This allowed a rapid expansion to Central and South America. Moreover, a second migration to North and Central America took place about 12,500 years ago, possibly via an interior route [Schurr and Sherry, 2004].

The different geographic location of HBV genotype F's subgroups and clusters could correlate with this double entrance to South America along the eastern (F1, cluster Ib) and western (F2, cluster II, III and IV) coasts. Where the continent becomes narrower or ends, these paths could have become closer resulting in the mixture of all clusters in Central America, as well as in the isolation of subgroups F1 (cluster Ib) and F2 (cluster IV) in Argentina [Campos et al., 2005].

By showing the highest similarity to genotype F, HBV genotype H has probably split off within the New World by early division of the progenitor HBV strains of the first settlers [Arauz-Ruiz et al., 2002]. However, as compared with the geographic location of genotype F, the distribution of the eighth HBV genotype within the American continent is limited as so far it seems to be restricted to Mexico and Central America.

HBV genotypes A (subgroup A2) and D are the signature of the European colonization that started in the sixteenth century and continued with two massive migrations during the end of the eighteenth century and the beginning of the nineteenth.

The African influx is also present in Latin America as a direct result of the slave trade during colonial times. Fine examples of this influence could be the predominance of genotype A (subgroup A1) in Brazil [Araujo et al., 2004] and its presence in northeastern Argentina [Campos et al., 2005], as well as the detection of genotype E in one Haitian pediatric patient living in Belgium [Liu et al., 2001] and two Dominican and Argentinean half-sisters of African descent [Mathet et al., 2006].

Strains ascribed to HBV genotypes B and C that were reported in Peru, Brazil and Argentina indicate the recent arrival of immigrants of Asian ethnicity.

As the worldwide origin of HBV genotype G – isolated in France and USA- remains to be unknown, the reported Mexican strains might be an import from USA due to the frequent migration between those countries.

2.3.1. HBV Genotypes in Mexico

Information regarding the molecular epidemiology of HBV in Mexico is still scanty.

The first study regarding the genetic characteristics of Mexican HBV strains revealed the predominance of divergent genotype F strains (66.6%), followed by genotype A -subgroup A2- (20%), D (6.7%) and G (6.7%) [Sánchez et al., 2002].

However, it was undertaken before the designation of HBV genotype H was even proposed to three strains found in Nicaragua and USA [Arauz-Ruiz et al., 2002]. Thus, some divergent strains formerly classified as HBV genotype F could be definitively ascribed to the genotype H.

A more recent report obtained an updated classification [Alvarado-Esquivel et al., 2006]. It concludes that at least 4 different genotypes circulate in Mexico, being HBV genotype H the most prevalent followed by genotypes G, A and D. Interestingly, the five strains ascribed to HBV genotype G in this study were detected as single isolates and not as genotype A-G mixed infections, as originally described [Kato et al., 2002].

It is important to remark that the frequency of genotype H among Mexican strains (75.5%) is the highest reported worldwide. Also, this particular genotype was found to be responsible for all HBV infections in the studied hemodialysis patients but only accounted for 50% of the infections among AIDS patients (the remaining half had been due to genotype G) [Alvarado-Esquivel et al., 2006]. HBV genotype H predominates in both acute or chronic hepatitis patients (74.0%) and among MSM (52.0%), while a noticeable frequency of genotype G (28.0%) was present only in the latter group, always contributing to mixed infections with genotypes A and H. Genotype D was detected in 32% of hepatitis patients [Sanchez et al., 2007].

2.3.2. Genotypes in Central America

In five different Central American countries (Guatemala, El Salvador, Honduras, Nicaragua and Costa Rica), the predominance of HBV genotype F (79.0%) was anticipated since this genotype is considered indigenous to the continent. Genotypes A, D and C (14.0%, 6.0% and 1.0%, respectively) were also present in serum samples from four groups of subjects (blood donors found HBsAg positive, patients with a diagnosis of acute or chronic hepatitis, HBsAg-positive pregnant women and Amerindian patients) [Arauz-Ruiz et al., 1997a].

In a later study [Arauz-Ruiz et al., 1997b], it was reported that two of the Nicaraguan and two of the Costa Rican strains previously ascribed to HBV genotype A clustered with isolates belonging to the A2 and A1 subgroups, respectively. Moreover, only one strain from Costa Rica belonged to genotype C and was closely related to strains from Vietnam, Laos and Bangladesh.

Interestingly, four Costa Rican isolates ascribed to HBV genotype D were more related to French strains associated with drug addiction encoding *ayw3* than to the Mediterranean strains encoding *ayw2* within this genotype. It has been proposed that at the time of the major Spanish immigration to Central America, genotype A might have dominated. However, since then, genotype D has been moving westward at the expense of genotype A as reflected by subtype distribution [Norder et al., 1993].

In general, the amino acid S gene sequence was quite conserved among all genotype F strains from Central America. While one of them encoded the subtype *ayw4*, the rest expressed the subtype *adw4,* thus exhibiting subtype heterogeneity among this HBV genotype.

On the other hand, two Nicaraguan strains -originally ascribed to genotype F- formed a divergent cluster from all other F strains. Subsequent sequencing of their complete genomes established that they represented a new genotype, for which the designation H was proposed [Arauz-Ruiz et al., 2002].

2.3.3. HBV Genotypes in Colombia

There are no reports regarding the molecular epidemiology of HBV infection in Colombia. However, only three HBV sequences from Colombia submitted to the GenBank were included as part of different study populations: two of them are ascribed to HBV genotype F [Norder et al., 1994 and 2004] and the remaining to genotype G [Toro et al., 2006].

2.3.4. HBV Genotypes in Venezuela

Confirming its indigenous origin, HBV genotype F prevails in different populations: Amerindians from western and southern Venezuela, hemodialysis patients, hemophiliacs, patients with chronic hepatitis, blood donors and pregnant women found positive for HBsAg [Blitz et al., 1998; Nakano et al., 2001]. Genotype A, D and B are also present.

Infections with multiple genotypes (but always involving genotype F) as well as two previously unrecognized geno-antigenic groups of HBV strain (subtype *ayw4* and *adw2*, genotype F) were detected among hemodialysis patients, who are considered high-risk patients for multiple infections by parentally transmitted viruses. The presence of these new associations was only found among patients attending the same hemodialysis Unit in Caracas, which suggests that nosocomial transmission could have been playing a role in viral dissemination. However, HBV genotype F/*ayw4* cannot exclusively be attributed to an outbreak in this

Venezuelan Unit, as the same association had previously been found in Central America [Arauz-Ruiz et al., 1997b].

Interestingly, a significant difference in prevalence of genotype F was found among blood donors screened as positive for anti-HBc and negative for HBsAg [Gutiérrez et al., 2004], as compared to the high proportion of this genotype observed among HBsAg positive blood donor samples collected during the same period (76.0% genotype F in HBsAg positive samples vs. 30.0% genotype F in HBsAg negative samples) [Devesa et al., 2004] or in the whole Venezuelan population [Blitz et al., 1998]. Occult infection isolates were mainly ascribed to genotypes A and D. Moreover, phylogenetic analysis of viral surface and core region revealed discrepancies in genotype assignment in some samples, suggesting the presence of mixed infections or genomic recombinations.

HBV genotype F and the co-infecting HDV genotype III, also indigenous to South America [Casey et al., 1996], are endemic in Amerindian population group from Venezuela and probably related to the outbreak of fulminant hepatitis with high mortality rate between 1979 and 1982 [Hadler et al., 1984; Torres et al., 1991; Nakano et al., 2001]. A more recent study revealed that HDV genotype I is also circulating among Venezuelan Amerindians, probably as a result of European immigrations, and is associated to HBV genotype F [Quintero et al., 2001].

When HBV strains from Venezuelan populations of African origin were analyzed, genotype F and A were equally distributed (50% each) [Quintero et al., 2002]. The higher prevalence of genotype A among these populations, compared to other Venezuelan groups, suggests that this genotype had been circulating in Africa in the 16th century and introduced in the Americas during the slave trade period. Several reports are in agreement with this hypothesis [Araujo et al., 2004]. Furthermore, the low degree of diversity observed among genotype A in Afro-Venezuelan populations supports the idea of a common introduction of these HBV isolates.

The Amerindian HBV genotype F is widely distributed among Amerindian and non-Amerindian groups (even in isolated communities). This might be due to a high rate of transmission of this genotype. However, another explanation for its circulation among the Afro-Venezuelan population is that this introduction might have occurred during the admixture of slaves with Amerindian groups.

The absence of the HBV genotype E in these communities is in agreement with recent phylogenetic studies that suggest a contemporary origin of this genotype in Africa [Takahashi et al., 2000].

2.3.5. HBV Genotypes in Brazil

Brazil can be differentiated from the other Latin American countries by the high prevalence of HBV genotypes A and D at the expense of genotype F. Molecular epidemiology studies carried out among blood donors as well as acute and chronic hepatitis patients from two Brazilian cities (Rio de Janeiro and São Paulo) revealed a predominance of HBV genotype A, followed by D and F strains. A similar distribution of genotypes was also observed among HIV positive patients [Sucupira et al., 2006].

Unexpectedly, in western Amazon, where HDV and HBV are highly prevalent causing outbreaks of fulminant hepatitis, the most prevalent HBV genotype is A and not F, as observed in other areas of the continent with HBV-HDV superinfection. HBV genotype D was not detected in the Brazilian Amazon [Viana et al., 2005].

The high proportion of isolates belonging to the Afro-Asian subgroup A1 suggests a common African origin for a large number of Brazilian HBV patients [Araujo et al., 2004]. Supporting this idea is the fact that HBV genotype A subgenotype A1 was present in all the studied samples from isolated Afro-Brazilian communities in central Brazil [Motta-Castro et al., 2005].

On the other hand, a different genotypic prevalence appears to exist among hemodialysis patients. In 1995, strains *adw2*/A and *ayw3*/D were equally predominant. However, a shift to a predominance of the latter has been observed since 1999, suggesting that subtype *ayw3* and genotype D are more likely to disseminate in the hemodialysis environment [Teles et al., 2002]. Since then, this hypothesis has been supported by several studies [Carrilho et al., 2004; Ferreira et al., 2006].

Only subsequent studies detected genotypes B and C which were first observed in São Paulo, probably due to the low percentage of Asian subjects included in previous studies [Sitnik et al., 2004].

Although there are many African descendants in Brazil due to slaves being brought from Africa from the 16th to the 19th centuries, HBV genotype E has not been detected yet.

2.3.6. HBV Genotypes in Peru

HBV Genotype F is the predominant genotype of the virus in the Peruvian Amazon Basin; it was reported -along with HDV genotype III- as the possible causes of acute hepatitis outbreaks occurred during 1992-1993 among Peruvian troops stationed at jungle outposts [Casey et al., 1996]. Strains from genotypes B and C were also observed [Campos et al., 2005].

2.3.7. HBV Genotypes in Bolivia

The only known report on HBV molecular epidemiology in Bolivia analyzed 6 strains ascribed to genotype F [Huy et al., 2006]. Phylogenetic analysis of the Bolivian HBV Pre-S gene sequences demonstrated the division of genotype F into four clusters -as previously reported [Mbayed et al., 2001; Devesa et al., 2004]- and revealed that the analyzed isolates were located in a cluster comprised of Argentinean strains (cluster IV of Mbayed). Moreover, full-length genome phylogenetic analysis on two of the 6 studied strains confirmed the existence of these four clusters with significant bootstrap values (98-100%). Therefore, the authors proposed to designate the four HBV genotype F clusters, previously included into 2 subgenotypes [Kramvis et al., 2005b], as subgenotypes F1, F2, F3 and F4 [Devesa et al., 2004]. This nomenclature is also currently accepted by other researchers [Schaefer S., 2007] and awaits final acceptance by the International Committee on Taxonomy of Viruses. Surprisingly, the so-called F4 cluster included an isolate which had been previously recognized as an F/C intergenotypic recombinant in a thorough review of the reported sequences [Simmonds and Midgley, 2005].

2.3.8. HBV Genotypes in Argentina

In support of the previous studies from other Latin American countries [Arauz-Ruiz et al., 1997; Blitz et al., 1998], the observed prevalence of HBV genotypes among Argentinean donors from blood banks were 64% for genotype F, 17.3% for each genotype A and D, and 1.3% for genotype C [França et al., 2004]. However, Buenos Aires and its metropolitan area were the only regions where genotype F did not predominate (31.8%). Its prevalence in the remaining studied geographical areas was equal or higher than 50%, the highest HBV genotype F prevalence being observed in northern Argentina (88.9%).

The calculated prevalence for genotypes A and D in the entire country is not reflected in the metropolitan area (22.7 and 40.9%, respectively), which mainly reflects the European immigrant origins of the Buenos Aires population.

Interestingly, the unique presence of HBV genotype A was reported in a pediatric population from Gualeguay city, a high-prevalence area which combines a low human population density with more emigration than immigration. The presence of strains ascribed to only one HBV genotype with very low nucleotide divergence and the absence of genotype F suggest a strong founder viral population [Mbayed et al., 1998].

It has been suggested that genotype D may have replaced genotype A in the Mediterranean area [Norder et al., 1993]. If this replacement had occurred after the spread of HBV strains in America, it could have explained the noticeable presence of genotype A in Argentina. Therefore, genotype D might have been introduced into Argentina by the more recent immigration from the Mediterranean area [Mbayed et al., 1998].

Unexpectedly, another study adds new information regarding the molecular epidemiology of HBV and the Americas as the mainly African-restricted HBV genotype E was detected in two sisters, one Argentinean and the other of Dominican origin. Both half-sisters had different fathers and exhibited an African DNA mitochondrial lineage [Mathet et al., 2006]. More importantly, one of them had been fully vaccinated and exhibited a significant anti-HBs antibody titer, despite of which a chronic hepatitis B was ongoing. This fact underscores the prime relevance to study the efficacy of the currently used HBV vaccines (subtype *adw*) to protect against genotype E HBV strains, since it is the most divergent among all genotypes within the *a* determinant (amino acids 107 to 147 of HBsAg). This feature, together with the emergence of S escape mutants, have raised concerns about the efficacy of the current vaccine in the African continent [Karthigesu et al., 1999].

Two reports have documented the prevalence of HBV genotype F, followed by genotypes A, D and B among chronically infected adults from the metropolitan area (Buenos Aires city and surroundings) [Telenta et al., 1997; López et al., 2002]. A more recent study agrees with these findings, establishing interesting associations, such as genotype F with HBeAg (+) chronic patients (44.8%) and severe fibrosis -Knodell Score: F3- (47.6%), and HBV genotype D with the HBeAg (-) population (45%) and mild fibrosis -Knodell Score F2- (38.6%) [Fainboim et al., 2006].

On the other hand, the genotypic prevalence among acute symptomatic HBV infected patients is slightly different; as F prevails (64.4%) followed by genotype A (31.1%) and D (4.4%). The low prevalence of the latter could be reflecting the progressive reduction of this HBV genotype in the area or its eventual association to asymptomatic forms of acute HBV hepatitis [Fainboim et al., 2006].

The HBV genotypic distribution in HIV/HBV coinfected patients shows a different picture as genotype A prevails (83.3%) and followed by D and F [Moretti et al., 2006]. Moreover, a high prevalence of HBV genotype A (approximately 90%) was also detected among street recruited IDUs (with and without HIV coinfection) from Buenos Aires city [Trinks et al., 2006]. Such value contrasts sharply with that registered among blood donors from the same city, which might suggest the dissimilar spread of HBV genotypes among different vulnerable groups of Buenos Aires inhabitants. Remarkably, an intergenotypic (D/A) recombination had been recently documented by the authors after a full-length HBV DNA sequence was analyzed from an isolate corresponding to an IDU patient (Trinks J. and Oubina J. R., unpublished data).

Chapter 3

HBV Variants and Mutants and Its Impact on the LAC Region

In an infected individual, HBV exists as a mixture of HBV quasispecies with a variant that is dominant among the others. The stability of a predominant HBV variant within the quasispecies pool is maintained by selection pressures from the host´s immune system, from viral factors which include overlapping open reading frames, viability and replication fitness of the virus, and from exogenous factors such as drugs including nucleoside/nucleotide analogs, interferon, and immune-based interventions (ie, HBIG and vaccination [Wai et al., 2004]). The accumulation of viral mutations depends on their generation rate and on the advantage granted to the resulting mutant virus [Brunetto et al., 1999].

Moreover, the structure of HBV genome with multiple overlapping ORFs, reduces the number of viable mutants and the rate of their production. However, selection and take-over of a mutant strain appear to be frequent events driven by both humoral and cellular host immune response and antiviral therapy. In addition, it was suggested that defective mutants may play an important role on HBV biology, interfering for example at replicative and transcriptional levels [Brunetto et al., 1999].

Along the HBV genome, a high number of mutations occur randomly. Some of these HBV variants may have a replication advantage over the others and may become dominant. Therefore, variants which replicate poorly, or are in some extent defective, fail to "survive", whereas those variants with enhanced survival have significant advantage in the viral population. The selection of one mutant over the others warrants a biological advantage to this mutant during its replication or a selective advantage to this mutant over wild-type virus in host-virus interactions [Brunetto et al., 1999].

From the pool of variants with similar replication potential, some will be positively selected by forces such as the humoral and cellular immune response; these variants are termed immune escape mutants, for example, HBsAg immune escape mutants [Carman et al., 1996].

Mutations may affect each of the ORFs identified within the HBV genome, and the genomic regulatory elements, enabling the virus to escape from selective pressures. Mutated S-genes allow the resulting mutant viruses to escape from both humoral and cellular host's immune response and reduce the effectiveness of diagnostics and immunoprophylaxis. As stated in the Introduction Section, escape mutants may be arbitrarily classified as immune escape mutants, diagnostic mutants, or as immune and diagnostic escape mutants. For example, these S-immune escape mutants are responsible for HBV-infections in successfully vaccinated persons. Mutated C-genes were found to provoke severe chronic liver diseases, and contribute strongly to immune escape at both the T helper and T cytotoxic level. Mutated X-genes can cause serious medical problems in blood donors by escaping the conventional test systems; and mutated P-genes were considered to be the reason for chemotherapeutic drug resistance [Kreutz et al., 2002]. It was also reported that usually, the enhancer II / core promoter mutant, the Pre-C stop codon mutant and the enhancer I / X promoter mutant are associated with severe and progressive liver disease. Due to the overlapping arrangement of the ORFs and regulatory elements on the HBV genome, it should be taken into account that almost every single mutation may influence more than one function of the corresponding nucleotide sequence [Pumpens et al., 2002].

In one study of HBV mutants, it was reported that in the Pre-C/C, Pre-S/S, Pol and X- genes, the prevalence of transversions is 20.6, 83.3, 50.0 and 20.0%; the prevalence of transitions is 44.0, 0.0%, 50.0 and 71.0%, respectively, while the prevalence of insertions is 14.7, 8.0, 0.0 and 0.0%, respectively. It was also observed that transitions are less frequently present in Pre-S/S-gene than in the other genes, and transversions are significantly more frequently in Pre-S/S-gene than in the other genes. With respect to the other types of mutations, no major differences have been found among their frequencies [Brunetto et al., 1999].

During the last years numerous mutations of the HBV genome have been described around the world, including the LAC region. Only replicative competent mutations which are associated with a functional change acquire clinical relevance. For example, a mutant virus may be the reason for HBV infections in successfully vaccinated persons, for failure of interferon or other antiviral therapy, and might be the origin of severe liver diseases such as chronic active hepatitis, liver cirrhosis and hepatocellular carcinoma [Kreutz et al., 2002].

3.1. Hepatitis B Virus Surface (HBsAg) Variants/Mutants

As stated in the Introduction Section, HBsAg variants/mutants may arise within or outside the *a* determinant (figure 2). Nucleotide substitutions in the corresponding coding region may not be necessarily reflected by aminoacidic changes within HBsAg nor in the (overlapped coded out-of-frame) Pol protein.

It was also suggested that areas upstream and downstream of the MHR may also be important in neutralization [Shields et al., 1999]. Mutations outside the MHR are frequent and tend to cluster in two regions around codons 44 to 49 and 152 to 213. The first region contains both a major histocompatibility complex class I (MHC-I)-restricted T-cell epitope and a B-cell epitope, whereas the second region, at least up to aa 207, exhibits both MHC-II T helper epitopes as well as B-cell epitopes. It was also reported that changes within this second region, located immediately downstream of the *a* determinant, may alter the conformation of this immunogenic determinant [Cuestas et al., 2006]. In agreement with this notion, Hou et al., showed that amino acid insertions and deletions in this region abolish the binding to anti-HBs antibodies [Hou et al., 1995]. Mutations in the HBsAg may result in amino acids substitutions, insertions, and/or deletions.

Hepatitis B surface "escape" mutants may arise in persons infected with HBV after they receive hepatitis B vaccine and/or HBIG, or may arise in chronic HBV patients during the natural course of HBV infection, due to the selective pressure of host's immune response for the survival of the fittest variant. A typical situation is the failure of passive/active hepatitis B immunization in newborns of HBV-infected mothers. According to Ogura et al., mutations within the *a* determinant during the natural course of infection are predominantly observed within the first loop (aa 107 to 138), whereas those induced under immune pressure due to active and/or passive immunization are more frequently observed within the second loop (aa 139 to 147) [Ogura et al., 1999].

HBsAg immune "escape" mutants would pose a substantial risk to the community, because current hepatitis B vaccines and HBIG are not effective in preventing infection with them. Moreover, blood units containing such mutants may potentially transmit HBV to recipients because some mutant viruses may escape detection by certain commercial HBsAg kits during a routine screening for this viral protein. It has been recently postulated that wild type HBsAg but not one of the main escape mutants (Gly145Arg) interacts intracellularly with endocytosed specific anti-HBs antibodies. This mechanism markedly reduces the hepatocyte secretion of the wild type virus simultaneously allowing the release of the escape mutant, suggesting a novel view for S mutants' emergence [Schilling et al., 2003].

3.1.1. Emergence of HBsAg "Escape" Mutants

The emergence of HBsAg "escape" mutants was first reported in an Italian boy in 1988 [Zanetti et al., 1988]. Since then, these "escape" mutants have been reported by many countries throughout the world. The far most important and best-documented mutation is the above mentioned substitution Gly145Arg. This mutant is stable over time, is able to replicate at high titer for many years and can be transmitted horizontally to other persons.

Other HBsAg "escape" mutants reported worlwide are those with amino acids substitutions at positions 120, 123, 124, 126, 129, 131, 133, 141 and 144 of the *a* determinant, insertion of 1, 2, 3 and 8 aa [Kreutz et al., 2002; Weber B., 2005; Coleman P. F., 2006] and deletions [Weinberger et al., 1999]. Other reported mutations that abolish the two loop structure of the *a* determinant that produce changes in the hydrophilicity, the electrical charge or the acidity of the loops are those at positions 127, 130, 134, 142, and 146. Furthermore, additional possible N-linked glycosylation sites (Gly130Asn, for example) or a change in the stability of a disulfide bridge (for example, Cys147Gly), could explain alterations of the two loop conformation [Kreutz et al., 2002].

HBsAg "escape" mutants have also been identified outside the *a* determinant, at positions 159, 183 and 184, that occur in either the vaccinated or unvaccinated population and which fail to bind properly to anti-HBs antibodies used in current commercial diagnostic kits for detection of HBsAg in the screening of hepatitis B disease [Oon et al., 1999; Coleman P. F., 2006]. Mutations which do not modify the *a* determinant but only change the subtypes alleles *w* to *r* (Lys 160Arg) do not have clinical importance. However, the subtype allelic mutation *d* to *y* (Lys122Arg) could be correlated with a higher failure rate of passive-active immunoprophylaxis in infants of HBeAg positive mothers. Mutations between codons 40 and 49, and between codons 198 and 208 that do not alter the *a* determinant were found in patients with HBIG prophylaxis after orthotopic liver transplantation. Those emerging within the first region could be selected by immune pressure because this region contains both a MHC-I-restricted T-cell epitope and a B-cell epitope, whereas the second region, exhibits both MHC-II T helper epitopes as well as B-cell epitopes [Kreutzer et al., 2002]. The clinical significance of S- mutants is still uncertain. However, the epidemiological importance of such mutants is supported by several reports.

3.1.2. HBsAg Mutants and Immune Prophylaxis

It was observed that in countries with massive vaccination programes, such as Taiwan, where a 10-fold decrease in the HBs carrier rate in children was successfully achieved, a substantial increase in HBs escape mutants of up to 28% was unfortunately documented [Pumpens et al., 2002]. Currently available HBV vaccines are made of viral sub-units containing HBsAg obtained either from plasma or by recombinant DNA technology. They are immunogenic, efficacious and safe. The efficacy of such vaccines against HBsAg "escape" mutants needs to be evaluated in order to determine whether vaccine modifications are required. It is likely that the inclusion of Pre-S2 and Pre-S1 epitopes in future HBV vaccines may eliminate or reduce the generation of vaccine associated "escape" mutants [Shouval et al., 2003], although for some authors the inclusion of Pre-S epitopes does not appear to increase the efficacy of the vaccine against S "escape" mutants. They suggest a vaccine containing the specific epitopes to induce neutralizing antibodies against the most common HBsAg mutants [Kuroda et al., 1991; Wilson et al., 1999].

In 1999, Wilson et al. developed an epidemiological-mathematical model of HBV in order to investigate the posible patterns of emergence of a vaccine-resistant strain. This model predicts that HBsAg "escape" mutants might become dominant over the wild-type after a putative period of at least 50 years, if conditions do favour the emergence of such mutants (e.g., high S-mutant infectiousness and low vaccine cross-immunity between wild-type and S-mutants strains) [Wilson et al., 1999].

3.1.3. HBsAg Mutants and Diagnosis

Another potential risk represented by such S-mutants, is that current HBsAg assays may be unable to properly detect all of them [Coleman P. F., 2006]. Thus, they could be potentially transmitted horizontally by blood transfusions of an asyntomatic carrier with no known risk factors of HBV infection, and whose HBsAg serology rendered a false negative result due to the harbouring of HBsAg mutant virus that escape detection by commercial HBsAg kits [Levicnic-Stezinar S., 2004; Thakur et al., 2005].

The sensitivity of HBsAg assays for S-mutants detection is continuously improved by the introduction of new HBsAg assays able to detect the so far described S-gene mutants. Polyclonal capture antibody-based assays do not guarantee full sensitivity, although they show better performance than most commercially available double monoclonal (capture and tracer) assays, which use antibodies against the second loop of the a determinant. The main drawback of using polyclonal antibodies immunoassays is that the specificity is lower than that of monoclonal assays. Mixtures of monoclonal antibodies able to recognize both wild-type an S-mutants strains may be used for HBsAg screening. In order to reduce the diagnostic failure related to HBsAg-negative mutants, it was suggested the tracer and capture antibodies against the Pre-S region should also be included in the diagnostic immunoassays. However, a potential risk exists that the analytical sensitivity for HBsAg of different genotypes even of wild-type virus could be decreased exists, if a mixture of antibodies is used. Antibodies directed against the second loop of the a determinant are not recommended for immunoassays since they fail to detect Gly145Arg mutants [Weber B., 2005].

Weber also suggested that for the screening of blood donors, anti-HBc testing should be used in combination with HBsAg, especially in those countries where HBV DNA detection by nucleic acid amplification technology is not mandatory [Weber, 2005], as it is the case in many parts of the LAC region.

3.1.4. HBsAg Mutants and Their Impact on the LAC Region

As stated above, some HBsAg mutants which either occur naturally or are selected through immunological pressure, have been shown to be clinically significant through their association with active and/or passive HBV immunoprophylactic failures and with loss of diagnostic accuracy. While sero-epidemiological data have been reported from some countries, data of HBsAg mutants in some areas of the world, as it is the case for Latin America, remain too poor, if not completely unknown. This kind of information is of crucial significance in an increasingly immunized world against HBV, where HBsAg mutants may not only become set up and spread within a population that is assumed to be protected, but may fail to be detected by immunodiagnostic kits. Thus, epidemiological studies from different parts of the world are required to understand the prevalence and characteristics of HBsAg mutants. When available, such information will be highly useful for both the future production of (a) new

HBV vaccine(s), and the design of new HBsAg detection immunoassays. This will also be of prime significance for avoiding false negative results in the screening of HBsAg in blood donors. Unfortunately, as stated above, there are scarce data about S- mutants/variants in the LAC region.

3.1.4.1. HBsAg Variants / Mutants in Mexico

One study from Mexico, showed a high degree of amino acid variability in the S protein of genotype F strains from 9 and 6 Mexican patients with chronic and acute hepatitis B infection, respectively. Seven of the nine strains isolated from the chronic carriers had at least one amino acidic substitution not previously reported outside the *a* determinant of the S protein; whereas only one of the 6 strains isolated from acute cases presented one unique amino acid substitution. The amino acidic changes observed among the different genotype F divergent strains were: Glu30Arg, Leu39Arg, Thr114Ser, Phe134Tyr, Tyr161Phe/Ser, Val180Ala, Pro214Leu, Leu216Stop, and Phe219Cys. One Mexican strain ascribed to genotype A had the Phe220Cys amino acid substitution. These amino acidic changes observed mainly outside the *a* determinant, did not identify any mutational hot spot of HBsAg, although the substitutions Tyr161 Phe/Ser and Val180Ala were found in two and three strains, respectively. Altered antigenicity due to amino acidic changes within the *a* determinant as well as outside this highly immunogenic region, have been described, although none of them are known as vaccine "escape" mutants. However, it is worth noticing that when S sequences from 2 genotype F strains exhibiting only the amino acid change Val180Ala were analyzed, one rendered a positive result for the detection of HBsAg, while the remaining proved negative. Furthermore, the S sequence from another genotype F strain that showed the amino acid substitutions Tyr161Phe and Val180Ala, also rendered a negative result for the detection of HBsAg.

Up to now, extensive studies relating some mutated S sequences ascribed to HBV genotype F with the decreased or absent reactivity of HBsAg with commercial current kits used for hepatitis B diagnosis have not been performed yet. Moreover, no data has already been recorded for genotype H strains. It might be possible -although not certainly asseverated- that some genotype amino acid substitutions such as Val180Ala, observed in two above mentioned HBsAg negative patients could have been responsible for the lack of detection of the viral surface antigen. It is noteworthy to mention that the S-mutant Val180Ala has also been identified in immunized Singapore infants [Oon et al., 1999].

The S protein mutation Leu216Stop, was observed in an isolated strain from an HBsAg positive chronic carrier. Its positivity indicates that the 10 N-terminal amino acids of the S protein are not essential for HBsAg detection using commercial immunoassays [Sánchez et al., 2002].

3.1.4.2. HBsAg Variants / Mutants in Venezuela

It has been reported that the HBV occult (defined also as residual) infections' prevalence in blood units, where anti-HBc antibodies are positive and HBsAg is negative, is approximately of 4.3% in this country. These residual infections are generally chronic, asymptomatic, and with low levels of viral replication. They are frequently associated with HBV variants that exhibit mutations in the surface gene that may affect the correct recognition by commercial tests, with Pre-C mutants and core internal deletions. In that study it was documented that although HBV ascribed to genotype F prevails among HBsAg positive samples from blood donors, residual infection isolates were mainly ascribed to genotypes A and D. The presence of mixed infection or recombination for some samples was also suggested. Amino acidic substitutions in the S protein, but not characteristic vaccine "escape" mutants, were detected. The most frequent S-mutants observed in residual infections of Venezuelan blood donors that might be responsible for the absence of HBsAg detection by routine assays were: Tyr100Cys and Cys220Phe [Gutiérrez et al., 2004].

Another group of investigators characterized HBV genotypes among 12 Yucpa Indians (figure 10), a population with highly endemic HBV infection. The characterized viruses characterized were all ascribed to genotype F, and in 2 individuals, a three-amino acidic novel deletion was identified just prior to the *a* determinant loop of the S protein. In spite of this, the amino acid sequences for the HBsAg protein from the Venezuelan strains were highly conserved, with only five positions of three strains containing the following amino acid changes: Leu9Pro, Leu98Val, Gln101Arg, Leu110Thr, and Pro111Asn. The samples in which they detected an S region deletion had no detectable HBsAg serological differences from the others [Nakano et al., 2001].

The circulation of an HBV isolate which exhibited point mutations within the S protein and could not be classified within any clade was documented in another study [Devesa et al, 2004]. This isolate (obtained from a blood donor), harboured the most famous, important and best-documented vaccine "escape" mutant: Gly145Arg. This isolate also presented a stop codon in the S protein (Leu94Stop), and the amino acid substitutions Cys137Trp and Cys139Tyr. Mutations of these cysteine residues are expected to be reflected as conformational changes of the major immunogenic HBsAg epitope, the *a* determinant. All these mutations suggest a strong HBV attempt to evade recognition by the host's immune system. It might have been expected that with such mutations, this mutated surface antigen could have been no longer recognized by specific antibodies used in current commercial immunoassays kits. However, this serum was positive for the detection of HBsAg, suggesting that a minor wild type HBV genome population could have contributed to the whole spectrum of circulating genomes. Another isolate ascribed to genotype F, from another blood donor harboured a point mutation in the initiation codon (A_TG/AC_G) of the HBsAg [Devesa et al., 2004].

3.1.4.3. HBsAg Variants / Mutants in Brazil

Sucupira et al., [2006] from Rio de Janeiro, Brazil, characterized some HBV strains from HIV-coinfected individuals treated with lamivudine as a component of the anti-retroviral therapy. Occult HBV infection was detected in 30% of the co-infected patients whose HBV serology was only positive for anti-HBc antibodies, and –unexpectedly- in 14% of such patients whose HBV serology only rendered positive results for both anti-HBs and anti-HBc antibodies. Possible explanations for the presence of HBV DNA in absence of HBsAg could have been related to: 1) low levels of circulating HBsAg; 2) the presence of S-gene variants, not detectable by current commercial immunoassays; 3) the presence of immune-complexes in which HBsAg is not available for its detection.

3.1.4.4. HBsAg Variants / Mutants in Argentina

Reports from Argentina have recently documented the appearance of HBsAg mutants despite the presence of usually protective anti-HBs antibodies [Mathet et al., 2003; Cuestas et al., 2006; Mathet et al., 2006; Mathet et al., 2007; López et al.,2007] as well as from cytotoxic T lymphocyte-specific clones [Cuestas et al., 2006]. Some of them occured naturally in HBV-chronically infective patients while others emerged under selected immunological pressure upon vaccination. So far, some of such mutants have not been reported in other regions of the world.

In 2003, Mathet et al. described the co-circulation of highly heterogeneous HBV quasispecies ascribed to genotype F showing S gene mutants in an Argentine patient who suffered from chronic hepatitis and cirrhosis, despite the fluctuacting detection of anti-HBs antibodies. In this study, several amino acid changes were detected inside and outside the *a* determinant of the HBsAg. Such changes might have resulted from the attempt of HBV to evade both the humoral and/or cellular immune response. Within this major determinant, the S-mutants Cys124Arg, Cys124Tyr, Cys138 Arg, Gln129Arg, Cys139Arg and Ser140Thr were observed. The cysteine residues present at positions 124, 138 and 139 are known to be important for appropriate folding of the S protein, while the cysteine residue present at position 69 is known to be needed for secretion of the HBsAg. Cys 90 and Cys176 are known to be dispensable residues. Outside the highly immunogenic *a* determinant, the following S-mutants were detected: Cys69Tyr, Cys90Tyr, Leu110Ile, Thr114Ala, Leu158Phe, Ala168Thr, Cys176Arg, and Gln178Ser/Asp [Mathet et al., 2003].

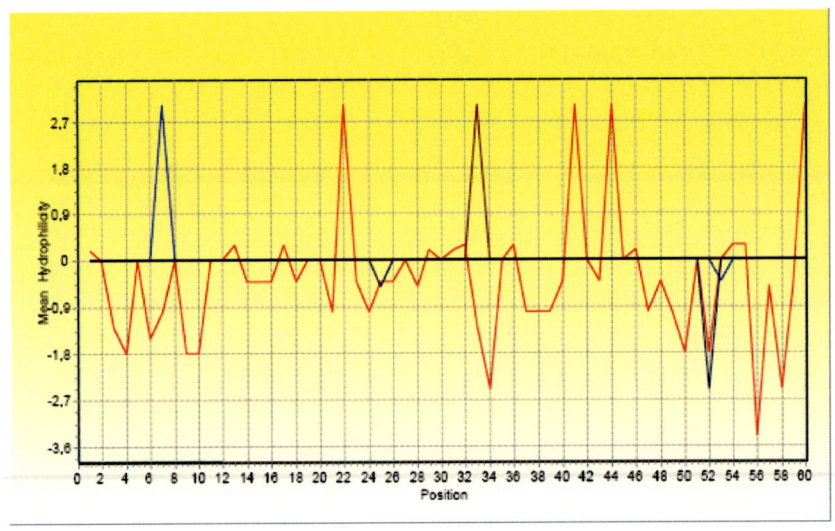

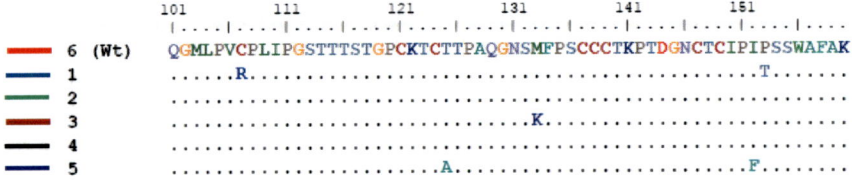

Figure 11. Hydrophylicity patterns obtained for HBV clones reprinted from J Clin Microbiol 44 (6): 2191-2198, Cuestas ML *et al*, "Unusual naturally occurring humoral and cellular mutated epitopes of hepatitis B virus in a chronically infected Argentine patient with anti-HBs antibodies", Copyright (2006), with permission from ASM. A partial analysis of the S protein (amino acid positions 101-160) encompassing the major hydrophilic region is shown. Amino acid sequence corresponding to the wild type (wt) sequence (as observed in clone 6) is drawn in a red pattern. Profiles depicting mutated clones are either shown in blue, green, brown, black or dark blue colors, and correspond to clones number 1, 2, 3, 4, or 5, respectively. Note the profound modification produced in the hydrophylicity pattern of clone 1 where a Cys107Arg substitution is observed. Likewise, a significant hydrophylicity change is observed at position 133 in clone 3.

Another study from Argentina, have also identified surface mutants/variants ascribed to genotype A, in a chronically infected Argentine patient who exhibited co-circulating anti-HBs antibodies. In this case, the amino acidic substitutions detected within the S protein were: Ser45Ala, Pro46His, Leu49His, Cys107Arg, Thr125Ala, Met133Lys, Ile152Phe, Pro153Thr, Tyr161Ser, Gly185Glu, Ala194Thr,Gly202Arg, and Ile213Leu [Cuestas et al., 2006]. Importantly, in some of the DNA clone-derived deduced amino acid sequences, crucial substitutions were associated with a change in the MHR hydrophilicity pattern (figure 11).

The detection of HBsAg mutants in two genotype E HBV isolates from two half-sisters exhibiting an African mitochondrial lineage was also reported in Argentina. Interestingly, the HBV-vaccinated sister exhibited anti-HBs co-cirulating antibodies without any previously known HBsAg "escape" mutant, while her HBV-unvaccinated half-sister showed an Asp144Ala HBsAg "escape" mutant without anti-HBs antibodies. Both sisters carried an unusual Leu209Val substitution within the S protein [Mathet et al., 2006]. As stated in the Introduction section, genotype E is the most divergent among all genotypes within the *a* determinant. It is of crucial significance to mention that the current HBV vaccine has the HBsAg genotype A2, subtype *adw2*, whereas genotype E isolates have subtype *ayw4*, and genotypes F isolates -widely distributed in Central and South America- have subtypes adw4q⁻ and/or ayw4. These subtype-epitopes are very immunogenic. This feature, together with the emergence of S "escape" mutants, have raised concerns about the efficacy of the current vaccine in the African continent where genotype E is the most predominant HBV genotype, and in Central and South America where F genotype is frequently observed.

3.1.4.5. Summary

Due to the frequency of HBsAg-mutants/variants in the LAC region and in other areas of the world, it may be prudent and important to adequate current commercial immunoassays kits used for hepatitis B diagnosis in order to detect individuals carrying such emerging viruses. This stresses the significance of epidemiological data in identifying current circulating S variants that give rise to phenotypically and antigenically distinct HBsAg particles. From a Public Health perspective, in addition to this, the epidemiological prevalence and clinical relevance of these HBsAg variants / mutants should be taken into consideration for the design and production of future successful vaccine(s) against HBV.

3.2. Hepatitis B Virus Basal Core Promoter (BCP) and Pre-C Variants/Mutants

The Pre-C/core gene encodes for two RNA transcripts: the pregenomic mRNA and Pre-C mRNA. Both regions are in the same reading frame but Pre-C gene starts at position 1814, whereas the core gene starts at 1901 (figure 12). The pregenomic mRNA is translated into the core protein (HBcAg) which is present within the virions. This mRNA also serves as a template for reverse transcription into the minus strand of HBV DNA. In contrast, the Pre-C mRNA translates into a Pre-C protein, which after processing at its C- and N-terminal ends, is secreted into the serum as HBeAg. This antigen is a non-structural protein believed to play a role as an immune tolerogen [Chen et al., 2005]. In the past, HBeAg seroconversion (loss of HBeAg with development of anti-HBe antibodies) has been assumed to result in clinical, biochemical and histologic remission of liver disease. However, subsequent studies have shown that progressive liver disease may develop in the absence of detectable HBeAg. Worldwide, as many as 30 % to 50 % of chronic HBV patients have HBeAg-negative liver disease.

Transcription of the Pre-C mRNA is under the control of several regulatory elements which include the basic core promoter which is placed between nucleotides 1744-1804. Mutations in this region decrease Pre-C mRNA transcription and subsequent HBeAg synthesis. The most common BCP variant involves a dual mutation A1762T / G1764A. These mutations have been proved to increase viral replication, possibly due to the inhibition of the Pre-core gene expression, thus explaining why they prevail over wild type virus during chronic infection. They are related to progression of chronic liver disease, and may be involved in the carcinogenesis of cirrhotic and non-cirrhotic patients [Liu et al., 2006]. These mutations are non-synonymous in the ORF of (out-of-frame) X gene, producing two amino acidic changes -Lys130Met and Val131Ile- discussed below (figure 12).

At the translational level, mutations within the Pre-C region can also block the translation of the Pre-C protein and subsequent production of HBeAg. The most common Pre-C variant is a point mutation G1896A (which results in a stop codon, instead of coding the amino acid tryptophan). This mutation is frequently accompanied by another point mutation (G to A) at nucleotide 1899. The frequency of the G1896A mutation varies for different genotypes, depending on the substitution of position 1858 opposing position 1896 in the stem of the Pre-C encapsidation signal or ε motif. The prevalence of the above mentioned mutation in the nucleotide 1896 among HBV isolates is presumably explained by the improved base pairing and stabilization of the stem-loop structure [Wai and Fontana, 2004].

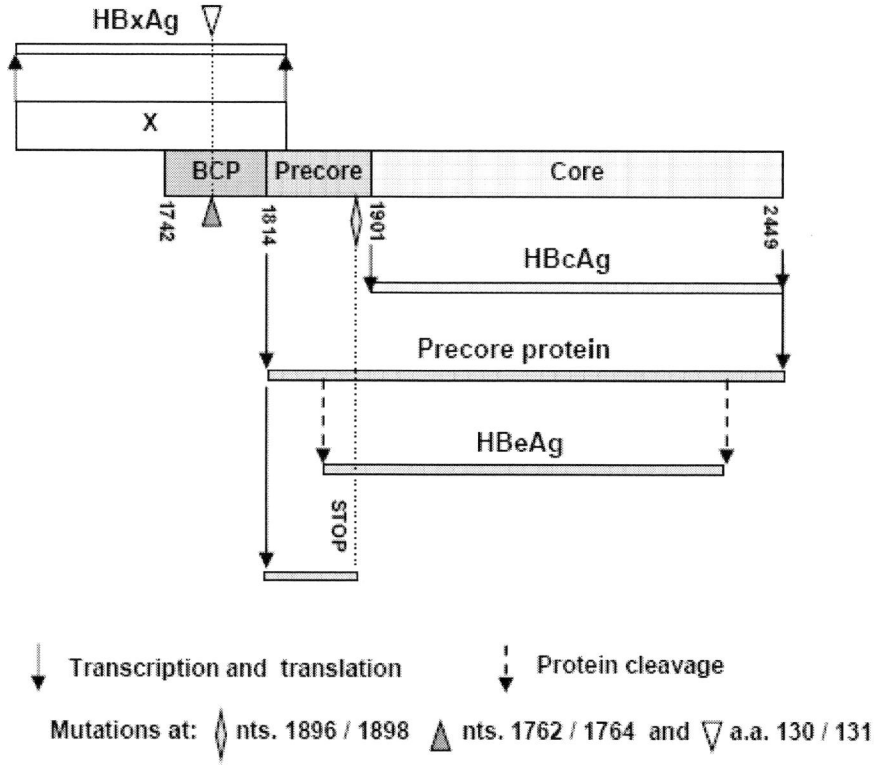

Figure 12. A) Schematic representation of the BCP (Basal core promoter), Pre-C and Core genomic regions and the overlapped (out-of-frame) X gene. The initiation of transcription and translation site is indicated with an arrow, and the viral proteins derived from these genes are shown: HBcAg, HBeAg, Pre-core and HBxAg. The HBeAg is a post-translational processing product of the Pre-core protein. The position of the protease cleavage sites within the Pre-core protein is indicated with an arrow. Mutations G1896A / G1899A can block the translation of the Pre-core protein and reduce the secretion of HBeAg. Other mutations are depicted. B) The HBV genome positions 1847-1907 are part of a pregenomic RNA encapsidation signal, consisting of a hairpin RNA structure. The dual mutations at positions 1858/1896, 1856/1898 are shown in blue circles. The AUG core start codon is depicted in red rectangle.

It is worth noticing that both BCP and Pre-core stop codon mutations are frequently observed in patients with advanced liver disease such as those with HCC [Baptista et al., 1999]. Other mutations, such as C1653T and T1753V (not T) mutants have also been associated with the progression of chronic hepatitis to cirrhosis and/or HCC [Takahashi et al., 1999].

As observed in figure 12, the overlapping nature of the X gene and the Pre-C region may be sometimes associated to nucleotidic changes that affect both of them. For example, frequent mutations as those observed at positions 1762 (A to T) and 1764 (G to A) may affect both the BCP region and simultaneously produce phenotypic changes in the X protein (Lys130Met and Val131Ile replacements, respectively). At present, it is not definitively established up to what extent mutations within such region affect the HBV gene expression, viral load or HBeAg status. However, a recent study reported that mutations within the Pre-core region such as G1896A, A1762T and G1764A enhance replication of (subgenotype) Bj clones in *in vitro* experiments. Moreover, fulminant hepatitis was frequent (13%) and associated with Bj subgenotype and lack of HBeAg, as well as high replication due to Pre-core mutations in patients with acute HBV infection [Ozasa et al., 2006].

As mentioned in precedent sections, Hepatitis B virus genotype F subdivides into two major subgenotypes, one made up mainly of strains from Central America (F1) and the other of strains from South America (F2). Norder et al. investigated the prevalence of mutation C1858T in the Pre-C region of genotype F from different geographical origins [Norder et al., 2003]. In such study, isolates of Central America (F1) exhibited the substitutions C1858T in the Pre-C region and Leu45Thr in the S gene; whereas in the remaining subgenotype, composed mainly of South American strains, all isolates had the nucleotide C at position 1858 in the Pre-C region and Leu at position 45 in the S gene. The lower nucleotide and amino acid divergences for S genes of Central American strains belonging to this clade may be due to the fairly recent introduction of this variant into the Hispanic population of Central America. Strains expressing C1858T are also present in Argentina and Brazil [Lopez et al., 2002; De Castro et al., 2001]. Although the S genes of these strains were not characterized, the S genes of three other strains from Argentina express Leu45Thr and cluster within subgenotype 1 [Mbayed et al., 2001] indicating its spread into South America.

Because of their frequent and worldwide prevalence, Pre-C and core promoter variants are believed to represent naturally occurring polymorphisms of the virus in humans.

3.2.1. Pre-C and BCP Variants / Mutants in Venezuela

Gutierrez et al. described a very low frequency of Pre-C mutations (10.5 %) in blood donors from Venezuela, who exhibited occult HBV infection [Gutierrez et al., 2004]. In the same way, Nakano et al., have not observed mutations in the

Pre-C region of genotype F HBV strains from Yucpa indians, a group with high endemicity HBV [Nakano et al., 2001]. In these studies synonymous and non-synonymous mutations in the Pre-C/core region were documented, as well as deletions near the N-terminal core region in the major B-cell epitope of HBcAg and HBeAg.

3.2.2. Pre-C and BCP Variants / Mutants in Brazil

Several reports suggested the infection by viral variants with Pre-C mutations in Brazil [De Castro et al., 2001; Rezende et al., 2005]. In this country, De Castro et al. reported a low frequency -in anti-HBe (+) patients- of both the stop codon mutation at nucleotide 1896 (24 %) and the double mutation in the basal core promotor 1762-1764 (20 %). In contrast -in the same sort of patients- another group described a high frequency (58.6 %) of mutation at position 1896 associated with a greater severity of liver disease [Rezende et al., 2005].

3.2.3. Pre-C and BCP Variants / Mutants in Argentina

In Argentina, the reports concerning both Pre-C and BCP mutations are dissimilar, according to the studied populations. França et al reported that HBV genotype F and mutant 1896 strains were predominant and strongly associated in a geographically broad Argentinean blood donor population [França et al., 2004]. However, Lopez et al. had previously demonstrated a high frequency of these mutations among genotype A and D HBV strains from patients with chronic hepatitis B. Moreover, the most prevalent genotype F HBV strains had been similarly recorded among HBeAg (+) and (-) patients [Lopez et al., 2002]. In an HBV chronically infected patient exhibiting chronic active hepatitis and cirrhosis despite the presence of anti-HBs antibodies, the BCP double mutant A1762T / G1764A was observed which suggests its involvement in the severe course recorded [Mathet et al., 2003; Mathet et al., 2007; Lopez et al., 2007; Mathet et al., unpublished].

3.3. HBx Variants / Mutants

Hepatitis B virus X protein (HBxAg) have a dual function, one related to its cytoplasmic localization, which can mediate the activation of signal transduction pathways, and another, a nuclear function, that may account for the interaction with transcription factors and components of the transcription apparatus to enhance the binding or activities of these proteins.

The HBxAg has been demonstrated to function as a transcriptional transactivator of a variety of viral and cellular promoter / enhancer elements. Although without binding directly to DNA, HBxAg can transactivate transcription through multiple cis-acting elements including AP-1, AP-2, ATF/CREB, NF-κB, C/EBP and HNF1 [Shih et al., 2000].

As mentioned above -due to the overlapping nature of the Pre-C/core and the X genes- the dual mutations A1762T and G1764A in the BCP are non-synonimous in the ORF of -out of frame- X gene, producing the amino acidic changes Lys130Met and Val131Ile (figure 12) [Poussin et al., 1999]. Furthermore, it was demonstrated that such changes enhance the binding of the nuclear factor HNF1 to the DNA, producing a highly transactivating effect [Li et al., 2002].

Kao et al. (Kao et al., 2003) documented that HBV chronic patients infected with these HBx variants have an increased risk of developing hepatocellular carcinoma.

Several reports from USA, as well as from several European and Asian countries have focused on these HBx mutations. In contrast, studies regarding these topics are almost lacking in Latin America and the Caribbean region.

3.3.1. HBx Variants / Mutants in Argentina

Lopez et al. have performed a longitudinal follow-up (3 years) during the course of an e-minus chronic hepatitis B infection in an F genotype-infected patient experiencing a severe disease outcome [Lopez et al., 2007]. In this study, HBx was the gene most commonly deleted; moreover, in 10 out of 13 deletant clones, this gene was completely absent. In several of the partial and full length X gene clones analyzed, the single mutation Lys130Met was documented. Interestingly, in the same blood samples, Mathet et al. observed the presence of HBV quasispecies with the dual mutation Lys130Met / Val131Ile (unpublished data). These ambiguities at the X gene sequence might be ascribed to the different set of primers used in each PCR amplification experiment and are indicative of the complexity of the genome population [Mathet et al., 2007].

3.4. Polymerase Gene Variants / Mutants

For many years, therapy of chronic hepatitis B has become one of the major clinical challenges for virologists and clinicians [Locarnini S., 2005]. Up to date, current monotherapy with conventional interferon alpha (IFN-α), and the nucleoside/nucleotide analogs -lamivudine (3TC or LMV) and adefovir dipivoxil (ADV)- remains unsatisfactory. Moreover, and in spite of the superiority of the recently licensed antiviral drugs such as peginterferon-α2a and the nucleoside analog entecavir (ETV) over the antiviral drugs mentioned above, more progress should be made in order to find out the most optimal treatment option (i.e., rescue of LMV, ADV or ETV resistance or for *de novo* treatment). At present, new antiviral drugs are under evaluation in experimental models and clinical trials. Among them, the most advanced is telbivudine (TBV), which has recently been licensed in Argentina.

Description of the own characteristics of the current antiviral drugs used in the treatment of chronic hepatitis B is beyond the scope of this chapter but below we are describing the most usual antiviral-resistant mutants, including those occurring naturally and those that are selected under antiviral drugs therapy as well as their epidemiology in the LAC region.

According to Wai et al., long term administration of LMV is associated with the emergence of drug-resistant mutants in 15% to 20%, 30% to 40%, and 60% of HBeAg-positive patients treated during 1, 2, or 4 years, respectively. The percentages for the appearance of LMV-resistant mutants in HBeAg-negative patients are even higher [Wai et al., 2004]. Removal of the drug leads to a reversion towards the wild-type virus because in the absence of antiviral pressure, wild-type strains originated from the very stable pool of viral cccDNA present in the nucleus of infected cells have the replication advantage [Zoulim F., 1999].

The four major patterns of HBV-resistant mutants selected during treatment with the antiviral drug LMV are:

- Leu528Met + Met552Val;
- Val521Leu + Leu528Met + Met552Val;
- Met552Ile;
- Leu528Met + Met552Ile;

Those mutations that occurred within the highly conserved YMDD motif of the C domain of the HBV polymerase, Met552Val/Met552Ile, are sufficient to confer resistance to LMV and other structurally related antivirals due to cross resistance among them. In addition to conferring drug resistance, the single mutations Met552Val and Met552Ile also reduce the *in vitro* replication efficiency of the virus in comparison with the wild-type (14% and 10% of HBV, respectively) [Ono et al., 2001]. According to a molecular model of HBV polymerase based on the crystal structure of the human immunodeficiency virus (HIV) reverse transcriptase (RT), it was suggested that the introduction of the β-methyl side chain of the amino acid valine or isoleucine, at position 552 of the RT domain of HBV polymerase, produces a steric hindrance that significantly affects the binding of LMV to the active site of the viral polymerase [Delaney et al., 2003]. The Met552Val mutation almost invariably occurs in combination with the Leu528Met, located in the B domain of the viral polymerase, whereas Met552Ile may occur alone or occasionally, in combination with the Leu528Met mutation [Wai et al., 2004]. *In vitro*, and in comparison with the single mutants Met552Ile and Met552Val, the double mutants Leu528Met + Met552Val, and Leu528Met + Met552Ile replicate better; however they are less replication fit than the wild-type virus (55 % and 68 % of the wild-type HBV, respectively) [Ono et al., 2001]. This means that the Leu528Met mutation rescues the defective replication fit of the single mutations Met552Val/Met552Ile. It was proposed that the Leu528Met mutation might compensate the impact of an amino acid substitution in the YMDD motif on viral replication [Fu et al., 1998] reducing the conformational imbalance caused by the single mutants Met552Ile or Met552Val.

In addition to enhancing the replication fitness of the YMDD mutants, the Leu528Met mutation renders them more resistant to LMV [Delaney et al., 2003].

A fourth well-characterized mutation in the HBV polymerase, that is associated with resistance to LMV, is the amino acid substitution Val521Leu. This mutation was only observed in combination with the double mutant Leu528Met + Met552Val suggesting that the Val521Leu mutation only emerges in HBV strains with this pattern of resistance to LMV. Interestingly, the Val521Leu mutant as a single mutation does not produce *in vitro* resistance to LMV, but in combination with Leu528Met + Met552Val, the resulting triple mutant Val521Leu + Leu528Met + Met552Val appears to be as resistant as the double mutant Leu528Met + Met552Val. Regarding the *in vitro* viral replication fitness, it was observed that the triple mutant Val521Leu + Leu528Met + Met552Val replicates an average of 90% with respect to the wild-type, thus, this triple mutant is more replication fit than the double mutant Leu528Met + Met552Val [Delaney et al., 2003]. This LMV-resistant triple mutant was found in 9% of the immunocompetent - LMV-resistant patients, and in 20% of LMV-resistant liver transplant or HIV-coinfected patients [Delaney et al., 2003].

Mutations within the HBV polymerase conferring drug resistance are also selected during treatment with ADV and ETV. Both antiviral drugs show a lower rate of resistance in comparison to LMV. As with LMV, the incidence of resistance to ADV or ETV increases with the duration of therapy.

Thus, ADV has a low resistance profile with 3%, 9%, 18%, and 28% after 2, 3, 4, or 5 years of treatment, respectively, whereas ETV has rarely produced resistance in naïve patients for up to 3 years [Tillmann H. L., 2007]. The resistance rate reported to ETV is 10% and 25% after 2 or 3 years, respectively, in patients with LMV failure, and 0.8% in naïve patients over 3 years [Tillmann H. L., 2007].The best-documented resistant mutants selected during treatment with ADV are the Asn584Thr and the Ala529Thr/Val mutant. These mutants develop in 5.9% of the patients within three years of therapy. Others less frequent ADV-resistant mutants are: Val432Met, Gln563Ser, Leu565Arg, Ser567Ala, Phe569Tyr, and Ile581Val. The double mutant Leu528Met + Ala529Val is resistant to both drugs, LMV and ADV, and are selected during combined antiviral therapy with them. Regarding the antiviral ETV, resistance to this drug was mainly observed in those LMV refractory patients that are receiving ETV as their new therapeutic regimen.

The best known resistant mutants selected during treatment with ETV on a background of LMV resistance mutations are: Ser532Gly, Ser550Ile, and Met598Val. ETV and ADV proved to be effective drugs in inhibiting replication of both wild-type and LMV-resistant HBV. Furthermore, ETV also demonstrated to have activity against ADV resistant mutants, while ADV seems to have activity against ETV resistant mutants which are also resistant to LMV.

3.4.1. Pol Mutants in the LAC Region as Compared with the Rest of the World

Emerging data from LAC countries will be discussed with those obtained in other world areas.

Resistance to antiviral therapy should be suspected in those patients undergoing an antiviral regimen for treatment of chronic hepatitis B, who suddenly present a virologic breakthrough accompanied by an increase in serum aminotransferase levels.

Although the introduction of antiviral therapy has led to the selection of several important polymerase mutants that confer resistance to most of the antiviral drugs currently used for the treatment of chronic hepatitis B, it should be taken into consideration that although naturally occurring polymerase gene mutations are very rare, they do exist. For example, the Ile581Val mutation that confers resistance to ADV, was identified in Germany in a patient chronically infected with HBV ascribed to genotype D, subgenotype D3, before ADV therapy was initiated. This naturally occurring ADV-resistant virus was also detected in HBV strains ascribed to genotype C, and in gibbons [Schildgen et al., 2006]. Other example of primary ADV resitance involved the HBV strains with the Leu565Arg mutation before starting ADV therapy. This natural mutant has been observed in most HBV strains ascribed to genotype A, subgenotype A2, which are prevalent in Europe and North America, but not in strains with other genotypes [Schildgen et al., 2006]. However, in Argentina, this naturally occurring ADV-resistant mutant was identified in patients chronically infected with HBV who did not undergo antiviral therapy with ADV. These isolated HBV strains that show natural resistance to ADV were ascribed to genotypes A, E and F (Cuestas M. L., unpublished data).

Naturally occurring LMV-resistant mutants were identified in two Mexican HBV isolates ascribed to genotype H. One of them showed the double mutant Leu528Met + Met552Val, whereas the other showed only the single mutant Leu528Met [Alvarado-Esquivel et al., 2006].

All these results suggest the clinical importance of testing for resistant mutants before antiviral therapy regimen is set up.

Regarding the organization of the HBV genome into ORFs, one interesting characteristic is that the P gene overlaps the other three viral genes (C, S and X; figure 1). The S-gene is completely overlapped by the polymerase gene, being the P ORF in the +1 reading frame in relation to the former. As a consequence, mutations in the S gene might produce changes in the overlapping polymerase gene and vice versa. It is of clinical importance the fact that the HBV S gene overlaps specifically the RT domain of the polymerase gene, because genetic alterations in the S gene selected as a consequence of immunotherapies such as HBV vaccination or HBIG, may not only affect the conformational structure of the HBsAg, but also the RT domain of Pol, which is an important target for the antiviral therapeutic against HBV. Thus, mutations that confer resistance to LMV, ADV o ETV, may appear as a result of the corresponding S gene mutations. In addition to this, a functionally significant alteration of the viral polymerase reflected as a lower replication activity of the RT, that may influence viral replication fitness, could be another possible result of such S gene mutations. However, polymerase function seems to be intact in most S mutant viruses detected worldwide [Protzer-Knolle et al., 1998]. Similarly, point mutations within the YMDD motif of the reverse transcriptase domain of the viral polymerase can change the aa-sequence of the S protein, but due to the downstream location of the YMDD motif respect to the antigenically relevant epitopes of HBsAg, these point mutations are neither correlated with significative changes of the antigenicity nor of the serotypes of HBsAg. However, the binding to anti-HBs antibodies may be reduced due to the alterations in the structure of HBsAg produced by the LMV-selected changes. It is unknown if LMV-selected HBsAg changes may produce viral variants that are not neutralized by vaccine induced anti-HBs antibodies, but it is possible that the widespread use of nucleoside/nucleotide analogs like LMV, ADV, ETV select S-mutants that have a reduced affinity for vaccine induced anti-HBs antibodies [Torresi J., 2002].

According to Torresi, the ability to modify a viral protein by mutations in an overlapping but unrelated ORF may produce HBV mutants with altered antigenicity and/or replication fitness and a natural history that might be different to wild-type HBV strains [Torresi J., 2002].

In one report from Brazil, it was documented that 70% of the patients under study with chronic hepatitis B and coinfected with HIV, had received LMV as antiviral therapy for 11 to 60 months. Forty four percent of them proved to be negative for seric HBsAg but have detectable HBV DNA (occult infection). Interestingly, all patients that showed LMV resistant mutations in HBV polymerase (30%), were HBsAg positive. In contrast, all patients with occult infection had absence of LMV resistant mutations. Among the mutations that conferrred resistance to LMV, the double mutant Leu528Met + Met552Val, as well as the triple mutant Val521Leu + Leu528Met + Met552Val were recorded. This triple mutant produced in the corresponding overlapped region of the S protein the amino acid substitutions Glu164Asp and Ile195Met, which showed reduced affinity to anti-HBs antibodies, in a similar way to the well-documented vaccine-induced escape mutants Gly145Arg [Sucupira et al., 2006].

Several residues of the HBV DNA polymerase fragment that overlaps with the ORF encoding HBsAg were affected by S-mutants identified in a chronically infected HBV Argentine patient with cocirculation of anti-HBs antibodies. Although the YMDD motif was conserved despite the presence of mutations within the S protein, several MHC-I and MHC-II restricted epitopes were affected. One of the identified mutations in the S protein was Met133Lys, that accounted for an amino acidic change at codon 489 which -in turn- resulted in the premature termination of the Pol protein. It is assumed that since certain mutations in the gene that encodes for viral polymerase may impair virus replication to a different extent, a minor population of intact genomes should be present to help the formation of viral particles by complementation [Mathet et al., 2003; Cuestas et al., 2006]. Moreover, *in vitro* studies have shown that the combination of wild type and mutated reverse transcriptase, yielded DNA levels (at late times) even higher than the wild type alone, suggesting that wild type polymerase might function in *trans* to boost RT mutated replication [Heipertz et al., 2007]

In summary, as a consequence of the organization of the HBV into ORFs, the selection of S mutants by the immune pressure can select variants with changes in the overlapping P gene coding for the viral polymerase, that may or may not confer resistance to antiviral therapy with nucleoside/nucleotide analogs. In a similar way, the selection of polymerase mutants during long-term therapy with nucleoside/nucleotide analogs, can select variants with changes in the overlapping S gene coding for the HBsAg, that may have a reduced binding to anti-HBs antibodies.

Chapter 4

Conclusions

HBV infection has special characteristics in LAC countries as it presents specific and practical challenges for its control. The endemic does not only bear a variable geographical and population distribution but the technical, economic, and Public Health resources are also different for each country.

Even though most of the ways of transmission and preventive strategies are known, the vaccination coverage is limited and the preventive programs are scant. In spite of the progress gained so far in the field of vaccine and therapy, HBV infection still remains a challenging health problem in LAC countries.

In addition to higher vaccination coverage, the simultaneous implementation of preventative measures involving both the individual risk behavior and the social, economic, and biologic vulnerability of some populations is essential. Among many adolescents and young adults, the individual behavior and/or the social vulnerability are dangerously combined for HBV acquisition, as dramatically observed among some local communities.

A comprehensive response to this endemic in the LAC region combining the utilization of available tools in terms of prevention, diagnosis, and treatment will undoubtedly help not only to diminish the number of new infections but also to reduce HBV morbi-mortality as well. The quality of health care coverage and the political commitment are key factors in the control of HBV.

Most of the LAC countries have the capacity not only to employ the existing knowledge properly but also to carry out research that could contribute to acquire a better understanding of the endemic in the region. Research in LAC countries could help to answer key questions about HBV infection. For instance, what is the influence of A to H genotypes in the transmission, in the pathogenicity, in the clinics, and in the prevention and diagnosis of HBV infection? Or what is the significance of escape mutants from diagnosis and/or natural or acquired immune response? Likewise, it will be important to investigate on the implication of variants/mutants within the F and H genotypes (which are common in LAC countries) both during the infection and progress to disease. Furthermore, it will be crucial to do research that help to identify the association between these genotypes and the emergence of resistant variants.

Not only basic viral and clinical research is needed on HBV infection in the LAC region. Many important questions about its epidemiology remain to be answered. An existing limitation is the scant knowledge about the actual incidence (and even prevalence for some local communities) rates of HBV transmission in the region as well as the different risk factors for its acquisition.

The impact of the utilization of the existing knowledge for the control of HBV infection in LAC countries has not yet met the desired expectations. Therefore, it is also important to conduct research in the sanitary systems of each country to help unveil the reasons accounting for failures in the implementation of strategies that have already proved to be effective, thus enabling health authorities to plan more efficacious health policies for the HBV control within the LAC countries.

The advent of Molecular Medicine has raised new expectations to achieve better diagnostics, prophylaxis and treatment. In the fight against chronic hepatitis B, many goals have been achieved, but the battle has not finished yet. There is still a long way to go, with new challenges to encounter. The future waits for new and important discoveries that will stand far beyond their intrinsic academic value since they will will surely bring about improvements to prevent and diagnose HBV infections and cure human beings.

Acknowledgments

This work was partially performed by using grants from CONICET (PIP 6065), UBA (UBACYT M057), and ANPCYT (PICT 10871). The authors are deeply grateful to Maria Mercedes Martinez for the design of figure 5 and to María Victoria Illas for enhancing readability.

References

[1] Aguiar, JI; Aguiar, E; Paniago, A; Cunha, R; Galvão, L; Daher, R. Prevalence of antibodies to hepatitis B core antigen in blood donors in the middle West region of Brazil. *Mem. Inst. Oswaldo Cruz*, 2001, 96 (2), 185-187.

[2] Aguiar, JI; de Souza, JA; Aguiar, ES; Oliveira, JM; de Lemos, ER; Yoshida, CF. Low prevalence of hepatitis B and C markers in a non-Amazonian indigenous population. *Braz. J. Infect. Dis*, 2002, 6 (5), 269-270.

[3] Allain, JP. Epidemiology of Hepatitis B virus and genotype. *J. Clin. Virol*, 2006, 36 (Suppl 1),S12-7.

[4] Allain, JP. Occult hepatitis B virus infection and transfusión. *J. Hepatol,* 2006, 44, 617-618.

[5] Alvarado-Esquivel, C; Sablon, E; Martínez-García, S; Estrada-Martínez, S. Hepatitis virus and HIV infections in inmates of a state correctional facility in Mexico. *Epidemiol. Infect*, 2005, 133 (4), 679-685.

[6] Alvarado-Esquivel, C; Sablon, E; Conde-González, CJ; Juárez-Figueroa, L; Ruiz-Maya, L; Aguilar-Benavides, S. Molecular analysis of hepatitis B virus isolates in Mexico: predominant circulation of hepatitis B virus genotype H. *World J. Gastroenterol,* 2006, 12 (40), 6540-6545.

[7] Alvarez-Muñoz, T; Bustamante-Calvillo, E; Martinez-Garcia, C; Moreno-Altamirando, L; Guiscafre-Gallardo, H; Guiscafre, JP; Muñoz, O. Seroepidemiology of the hepatitis B and delta in the southeast of Chiapas, Mexico. *Arch. Invest. Med. (Mex)*, 1989, 20 (2), 189-195.

[8] Alvarez-Muñoz, MT; Bustamante-Calvillo, ME; Guiscafre-Gallardo, JP; Muñoz, O. Hepatitis B and delta: the prevalence of seroepidemiological

markers in volunteer blood donors and their families. *Gac. Med. Mex*, 1991, 127 (5), 399-404.

[9] Andrade, ZA; Lesbordes, JL; Ravisse, P; Parana, R; Prata, A; Barberino, JS; Trepo, C. Fulminant hepatitis with microvesicular steatosis (a histologic comparison of cases occurring in Brazil-Labrea hepatitis-and in central Africa-Bangui hepatitis). *Rev. Soc. Bras. Med. Trop*, 1992, 25 (3), 155-160.

[10] Andrade, AF; Oliveira-Silva, M; Silva, SG; Motta, IJ; Bonvicino, CR. Seroprevalence of hepatitis B and C virus markers among blood donors in Rio de Janeiro, Brazil, 1998-2005. *Mem. Inst. Oswaldo. Cruz*, 2006, 101 (6), 673-676.

[11] Antoni, BA; Rodriguez-Crespo, I; Gómez-Gutiérrez, J; Nieto, M; Peterson, D; Gavilanes, F. Site-directed mutagenesis of cysteine residues of hepatitis B surface antigen. Analysis of two single mutants and double mutant. *Eur. J. Biochem*, 1994, 222 (1), 121-7.

[12] Araujo, NM; Mello, FC; Yoshida, CF; Niel, C; Gomes, SA. High proportion of subgroup A' (genotype A) among Brazilian isolates of Hepatitis B virus. *Arch. Virol*, 2004, 149 (7), 1383-1395.

[13] Arauz-Ruiz, P; Norder, H; Visona, KA; Magnius, LO. Genotype F prevails in HBV infected patients of hispanic origin in Central America and may carry the precore stop mutant. *J. Med. Virol*, 1997, 51 (4), 305-312.

[14] Arauz-Ruiz, P; Norder, H; Visona, KA; Magnius, LO. Molecular epidemiology of hepatitis B virus in Central America reflected in the genetic variability of the small S gene. *J. Infect. Dis*, 1997, 176 (4), 851-8.

[15] Arauz-Ruiz, P; Norder, H; Robertson BH; Magnius, LO. Genotype H: a new Amerindian genotype of hepatitis B virus revealed in Central America. *J. Gen. Virol*, 2002, 83 (Pt 8), 2059-2073.

[16] Ballester, JM; Rivero, RA; Villaescusa, R; Merlin, JC; Arce, AA; Castillo, D; Lam, RM; Ballester, A; Almaguer, M; Melians, SM; Aparicio, JL. Hepatitis C virus antibodies and other markers of blood-transfusion-transmitted infection in multi-transfused Cuban patients. *J. Clin. Virol*, 2005, 34 (Suppl 2), S39-46.

[17] Bancroft, WH; Snitbhan, R; Scott, RM; Tingpalapong, M; Watson, WT; Tanticharoenyos, P; Karwacki, JJ; Srimarut, S. Transmission of hepatitis B virus to gibbons by exposure to human saliva containing hepatitis B surface antigen. *J. Infect. Dis*, 1977, 135 (1), 79-85.

[18] Bartholomeusz, A; Schaefer, S. Hepatitis B virus genotypes: comparison of genotyping methods. *Rev. Med. Virol*, 2004, 14(1), 3-16.

[19] Baptista, M; Kramvis, A; Kew, MC. High prevalenc of 1762 (T) 1764 (A) mutations in the basic core promoter of hepatitis B virus isolated from black

Africans with hepatocellular carcinoma compard with asymptomatic carriers. *Hepatology,* 1999, 29 (3), 946-953.

[20] Bensabath, G; Hadler, SC; Soares, MC; Fields, H; Maynard, JE. Epidemiologic and serologic studies of acute viral hepatitis in Brasil's Amazon Basin. *Bull. Pan. Am. Health Organ*, 1987, 21 (1), 16-27.

[21] Blitz-Dorfman, L; Monsalve, F; Porto, L; Weir, J; Arteaga, M; Padron, G; León, P; Echevarria, JM. Epidemiology of hepatitis C virus in western Venezuela: lack of specific antibody in Indian communities. *J. Med. Virol*, 1994, 43 (3), 287-290.

[22] Blitz, L; Pujol, FH; Swenson, PD; Porto, L; Atencio, R; Araujo, M ; Costa, L; Monsalve, DC; Torres, JR; Fields, HA; Lambert, S; Van Geyt, C; Norder, H; Magnius, LO; Echevarría, JM; Stuyver, L. Antigenic diversity of hepatitis B virus strains of Genotype F in Amerindians and other population groups from Venezuela. *J. Clin. Microbiol*, 1998, 36 (3), 648-651.

[23] Bollyky, PL; Rambaut, A; Harvey, PH; Holmes, EC. Recombination between sequences of hepatitis B virus from different genotypes. *J. Mol. Evol*, 1996, 42 (2), 97-102.

[24] Boulos, R; Ruff, AJ; Nahmias, A; Holt, E; Harrison, L; Magder, L; Wiktor, SZ; Quinn, TC; Margolis, H; Halsey, NA. Herpes simplex virus type 2 infection, syphilis, and hepatitis B virus infection in Haitian women with human immunodeficiency virus type 1 and human T lymphotropic virus type I infections. The John Hopkins University (JHU)/Centre pour le Developpement et la Sante (CDS) HIV Study Group. *J. Infect. Dis*, 1992, 166 (2), 418-420.

[25] Bowyer, SM; Sim, JG. Relationships within and between genotypes of hepatitis B at points across the genome: footprints of recombination in certain isolates. *J. Gen. Virol*, 2000, 81(Pt 2), 379-92.

[26] Brunetto, MR; Rodriguez, UA; Bonino, F. Hepatitis B virus mutants. *Intervirology,* 1999, 42 (2-3), 69-80.

[27] Buitrago, B ; Hadler, SC ; Popper, H ; Thung, SN ; Gerber, MA ; Purcell, RH ; Maynard, JE. Epidemiologic aspects of Santa Marta hepatitis over a 40-year period. *Hepatology*, 1986, 6 (6), 1292-1296.

[28] Burda, MR ; Günther, S ; Dandri, M ; Will, H ; Petersen, J. Structural and functional heterogeneity of naturally occurring hepatitis B virus variants. *Antiviral Res*, 2001, 52 (2), 125-138.

[29] Busek, SU ; Babá, EH ; Tavares Filho, HA ; Pimenta, L ; Salomão, A ; Correa-Oliveira, R ; Oliveira, GC. Hepatitis C and hepatitis B virus infection in different hemodialysis units in Belo Horizonte, Minas Gerais, Brazil. *Mem. Inst. Oswaldo Cruz*, 2002, 97 (6), 775-778.

[30] Cabezas, C ; Gotuzzo, E ; Escamilla, J ; Phillips, I. Prevalence of serological markers of viral hepatitis A, B and delta in apparently healthy schoolchildren of Huanta, Peru. *Rev. Gastroenterol. Peru*, 1994, 14 (2), 123-134.

[31] Cabrerizo, M.; Bartolomé, J.; Caramelo, C.; Barril, G.; Carreño, V. Molecular analisis of hepatitis B virus DNA in serum and peripheral blood mononuclear cells from hepatitis B surface antigen-negative cases. *Hepatology*, 2000, 32 (1), 116-123.

[32] Cacciola, I; Pollicino, T; Squadrito, G; Cerenzia, G; Orlando, ME; Raimondo, G. Occult hepatitis B virus infection in patients with chronic hepatitis C liver disease. *N. Engl. J. Med*, 1999, 341 (1), 22-26.

[33] Campos, RH; Mbayed, VA; Pineiro Y Leone, F. Molecular epidemiology of hepatitis B virus in Latin America. *J. Clin. Virol,* 2005, 34 (Suppl 2), S8-S13.

[34] Carman, WF; Jacyna, MR; Hadziyannis, S; Karayiannis, P; McGarvey, MJ, Makris, A; Thomas, HC. Mutation preventing formation of hepatitis B e antigen in patients with chronic hepatitis B infection. *Lancet*, 1989, 2 (8663), 588-591.

[35] Carman, WF; Trautwein, C; van Deursen, FJ; Colman, K; Dornan, E; McIntyre, G; Waters, J; Kliem, V; Muller, R; Thomas, HC; Manns, MP. Hepatitis B virus envelope variation after transplantation with and without hepatitis B immune globulin prophylaxis. *Hepatology*, 1996, 24 (3), 489-93.

[36] Carreto-Vélez, MA; Carrada-Bravo, T; Martínez-Magdaleno, A. Seroprevalence of HBV, HCV and HIV among blood donors in Irapuato, Mexico. *Salud Pública Mex*, 2002, 45 (Suppl 5), S690-693.

[37] Carrilho, FJ; Moraes, CR; Pinho, JR; Mello, IM; Bertolini, DA; Lemos, MF; Moreira, RC; Bassit, LC; Cardoso, RA; Ribeiro-dos-Santos, G; Da Silva, LC. Hepatitis B virus infection in Haemodialysis Centres from Santa Catarina State, Southern Brazil. Predictive risk factors for infection and molecular epidemiology. *BMC Public Health*, 2004, 27 (4), 13.

[38] Casey, JL; Niro, GA; Engle, RE; Vega, A; Gomez, H; McCarthy, M; Watts, DM; Hyams, KC; Gerin, JL. Hepatitis B virus (HBV)/hepatitis D virus (HDV) coinfection in outbreaks of acute hepatitis in the Peruvian Amazon basin: the roles of HDV genotype III and HBV genotype F. *J. Infect. Dis,* 1996, 174 (5), 920-6.

[39] Chen, BF; Liu, CJ; Jow, GM; Chen, PJ; Kao, JH; Chen DS. Evolution of Hepatitis B virus in an acute hepatitis B patient co-infected with genotypes B and C. *J. Gen Virol*, 2006, 87 (Pt 1), 39-49.

[40] Chen, M; Sallberg, M; Hughes, J; Jones, J; Guidotti, LG; Chisari, FV; Billaud, JN; Milich, DR. Immune tolerance split between hepatitis B virus precore and core proteins. *J. Virol*, 2005, 79, 3016-3027.

[41] Chu, CJ; Hussain, M; Lok, AS. Hepatitis B virus genotype B is associated with earlier HBeAg seroconversion compared with hepatitis B virus genotype C. *Gastroenterology*, 2002, 122 (7), 1756-1762.

[42] Chudy, M; Schmidt, M; Czudai, V; Scheiblauer, H; Nick, S, Mosebach, M; Hourfar, MK, Seifried, E; Roth, WK; Grunelt, E; Nubling, CM. Hepatitis B virus genotype G monoinfection and its transmission by blood components. *Hepatology*, 2006, 44 (1), 99-107.

[43] Coleman, PF.Detecting hepatitis B surface antigen mutants. *Emerg. Infect. Dis*, 2006, 12 (2), 198-203.

[44] Couroucé, AM; Holland, PV; Muller, JY; Soulier, JP. HBs antigen subtypes. *Bibl. Haematol*, 1976, 42, 1.

[45] Craig, PG; Bryan, JP; Miller, RE; Reyes, L; Hakre, S; Jaramillo, R; Krieg, RE. The prevalence of hepatitis A, B and C infection among different ethnic groups in Belize. *Am. J. Trop. Med. Hyg*, 1993, 49 (4), 430-434.

[46] Cruz, JR; Perez-Rosales, MD; Zicker, F; Schmunis, GA. Safety of blood supply in the Caribbean countries: role of screening blood donors for markers of hepatitis B and C viruses. *J. Clin. Virol*, 2005, 34 (Suppl 2), S75-80.

[47] Cuestas, ML; Mathet, VL; Ruiz, V; Minassian, ML; Rivero, C; Sala, A; Corach, D; Alessio, A; Pozzati, M; Frider, B; Oubiña, JR. Unusual naturally occurring humoral and cellular mutated epitopes of hepatitis B virus in chronically infected argentine patient with anti-HBs antibodies. *J. Clin. Microbiol*, 2006, 44 (6), 2191-8.

[48] Cui, C; Shi, J; Hui, L; Xi, H; Zhuoma; Quni; Tsedan; Hu, G. The dominant hepatitis B virus genotype identified in Tibet is a C/D hybrid. *J Gen Virol*, 2002, 83 (Pt 11), 2773-7

[49] Dane, DS; Cameron, CH; Briggs M. Virus-like particles in serum of patients with Australia-antigen-associated hepatitis. *Lancet,* 1970, 1(7649), 695-698.

[50] de Almeida Pereira, RA; Mussi, AD; de Azevedo e Silva, VC; Souto, FJ. Hepatitis B Virus infection in HIV-positive population in Brazil: results of a survey in the state of Mato Grosso and a comparative analysis with other regions of Brazil. *BMC Infect. Dis*, 2006, 25 (6), 34.

[51] De Castro, L; Niel, C; Gomes, SA. Low frequency of mutations in the core promoter and precore regions of hepatits B virus in anti-HBe positive Brazilian carriers. *BMC Microbiol*, 2001, 1, 10.

[52] de Franchis, R; Hadengue, A; Lau, G; Lavanchy, D; Lok, A; McIntyre, N; Mele, A; Paumgartner, G; Pietrangelo, A; Rodes, J; Rosenberg, W; Valla, D; EASL Jury. EASL International Consensus Conference on Hepatitis B. *J. Hepatol*, 2003, 39 (Suppl 1), S3-25.

[53] Delaney, WE 4th; Yang, H; Westland, CE; Das, K; Arnold, E; Gibbs, CS; Miller, MD; Xiong, S. The hepatitis B virus polymerase mutation rtV173L is selected during lamivudine therapy and enhances viral replication in vitro. *J. Virol*, 2003, 77 (21), 11833-41.

[54] de Marquez, ML; Galindez, E; Camacho, G; Carvajal, R; Mata, M; Tombazzi, C; Castro, E; Marques, F; Escudero, J; Lecuna, V. Epidemiology of viral hepatitis in Venezuela: preliminary results of phase 1. Prevalence in the metropolitan area. *GEN*, 1993, 47 (4), 215-220.

[55] de Paula, VS; Arruda, ME; Vitral, CL; Gaspar, AMC. Seroprevalence of viral hepatitis in riverine communities from the western region of the Brazilian Amazon Basin. *Mem. Inst. Oswaldo Cruz*, 2001, 96 (8), 1123-1128.

[56] Devesa, M; Rodríguez, C; León, G; Liprandi, F; Pujol, FH. Clade analysis and surface antigen polymorphism of hepatitis B virus American genotypes. *J. Med. Virol*, 2004, 72 (3), 377-384.

[57] Ding, X; Mizokami, M; Yao, G; Xu, B; Orito, E; Ueda, R; Nakanishi, M. Hepatitis B virus genotype distribution among chronic hepatitis B virus carriers in Shanghai, China. *Intervirology*, 2001, 44 (1), 43-47.

[58] Dryden, KA; Wieland, SF; Whitten-Bauer, C; Gerin, JL; Chisari, FV; Yeager M. Native hepatitis B virions and capsids visualized by electron cryomicroscopy. *Mol. Cell*, 2006, 22 (6), 843-850.

[59] Dutra Souto, FJ; Fernandes Fontes, CJ; Coimbre Gaspar, AM. Outbreak of hepatitis B virus in recent arrivals to the Brazilian Amazon. *J. Med. Virol*, 1998, 56 (1), 4-9.

[60] Eble, BE; MacRae, DR; Lingappa, VR; Ganem, D. Multiple topogenic sequences determine the transmembrane orientation of hepatitis B surface antigen. *Mol. Cell Biol*, 1987, 7 (10), 3591-3601.

[61] Echevarría, JM; Leon, P. Epidemiology of viruses causing chronic hepatitis among populations from the Amazon Basin and related ecosystems. *Cad. Saude Publica*, 2003, 19 (6), 1583-1591.

[62] Echevarría, JM; Avellon, A. Hepatitis B virus genetic diversity. *J. Med. Virol*, 2006, 78 (Suppl 1), S36-42.

[63] Erhardt, A; Reineke, U; Blondin, D; Gerlich, WH; Adams, O; Heintges, T; Niederau, C; Haussinger, D. Mutations of the core promoter and response to

interferon treatment in chronic replicative hepatitis B. *Hepatology*, 2000, 31 (3), 716-725.

[64] Fainboim, H; Gonzalez, J; Fassio, E; Martinez, A; Otegui, L; Eposto, M; Cahn, P; Marino, R; Landeira, G; Suaya, G; Gancedo, E; Castro, R; Brajterman, L; Laplume, H. Prevalence of hepatitis viruses in an anti-human immunodeficiency virus-positive population from Argentina. A multicentre study. *J Viral Hepat*, 1999, 6 (1), 53-7.

[65] Fainboim, HA, Pezzano, S; Bouzas, M; Alvarez, E; Schroder, T; Fernandez Giuliano, S; Campos, R; Mbayed, V. Genotipos de HBV en Buenos Aires. Argentina: Distribución y posible significado en diferentes situaciones clínicas. *Acta Gastroenterol. Latinoam*, 2006, 36 (Supl. 3), S62.

[66] Fares, MA; Holmes, EC. A revised evolutionary history of hepatitis B virus (HBV). *J. Mol. Evol*, 2002, 54 (6), 807-14.

[67] Fay, OH; Hadler, SC; Maynard, JE; Pinheiro, FP. Hepatitis in the Americas. *Bull PAHO*, 1985, 6, 1.

[68] Fay, OH. Hepatitis B in Latin America: epidemiological patterns and eradication strategy. The Latin American Regional Study Group. *Vaccine*, 1990, 8 Suppl, S100-6.

[69] Ferreira, RC; Teles, SA; Dias, MA; Tavares, VR; Silva, SA; Gomes, SA; Yoshida, CF; Martins, RM. Hepatitis B virus infection profile in hemodyalisis patients in Central Brazil: prevalence, risk factors, and genotypes. *Mem. Inst. Oswaldo Cruz*, 2006, 101 (6), 689-692.

[70] Figueiredo, JFC; Machado, AA; Martinez, R; et al. Prevalências das infeções pelos vírus, anticorpos e hepatites B e C na Reserva Indígena Xacriabá, MG, Brasil. *Rev. Soc. Bras. Med. Trop*, 2000, 33, 211.

[71] França, PHC; González, JE; Munné, MS; Brandão, LH ; Gouvea, VS ; Sablon, E ; Vanderborght, BOM. Strong association between genotype F and hepatitis B Virus (HBV) e antigen-negative variants among HBV-infected argentinean blood donors. *J. Clin. Microbiol*, 2004, 42 (11), 5015-5021.

[72] Fu, L; Cheng, YC. Role of additional mutations outside the YMDD motif of the hepatitis B virus polymerase in L(-)SddC (3TC) resistance. *Biochem. Pharmacol*, 1998, 55 (10), 1567-72.

[73] Fujiwara, K; Tanaka, Y; Paulon, E; Orito, E; Sugiyama, M; Ito, K; Ueda, R; Mizokami, M; Naoumov, NV. Novel type of hepatitis B virus mutation: replacement mutation involving a hepatocyte nuclear factor 1 binding site tandem repeat in chronic hepatitis B virus genotype E. *J. Virol,* 2005, 79(22), 14404-14410.

[74] Ganem, D; Prince, AM. Hepatitis B virus infection - natural history and clinical consequences. *N. Engl. J. Med*, 2004, 350, 1118-1129.

[75] Garay, ME; Marina, G; Martinez, M; Berchan Perera, M; Ramia, R; Falco, A; Olivo, L; Fainboim, H. Prevalencia intermedia de infección por virus B en la Provincia de Salta Argentina. *Acta Gastroenterol. Latinoam*, 2006, 36 (Supl.3), S63.

[76] Garcia-Montalvo, BM; Farfan-Ale, JA; Acosta-Viana, KY; Puerto-Manzano, FI. Hepatitis B virus DNA in blood donors with anti-HBc as a possible indicator of active hepatitis B virus infection in Yucatan, Mexico. *Transfus. Med*, 2005, 15 (5), 371-378.

[77] Georgi-Geisberger, P; Berns, H; Loncarevic, IF; Yu, ZY; Tang, ZY; Zentgraf, H; Schroder, CH. Mutations on free and integrated hepatitis B virus DNA in a hepatocellular carcinoma: footprints of homologous recombination. *Oncology*, 1992, 49 (5), 386-95

[78] Gish, RG; Gadano, AC. Chronic hepatitis B: current epidemiology in the Americas and implications for management. *J. Viral. Hepat*, 2006, 13 (12), 787-798.

[79] Guidotti, LG; Rochford, R; Chung, J; Shapiro, M; Purcell, R; Chisari, FV. Viral clearance without destruction of infected cells during acute HBV infection. *Science*, 1999, 284 (5415), 825-9.

[80] Günther S. Genetic variation in HBV infection: genotypes and mutants. *J. Clin. Virol*, 2006, 36 (Suppl 1):S3-S11.

[81] Gutiérrez, C; Leon, G; Loureiro, CL; Uzcategui, N; Liprandi, F; Pujol, FH. Hepatitis B virus DNA in blood samples positive for antibodies to core antigen and negative for surface antigen. *Clin. Diagn. Lab. Immunol*, 1999, 6 (5), 768-70.

[82] Gutiérrez, C; León, G; Liprandi, F; Pujol, FH. Low impact of silent hepatitis B virus infection on the incidence of post-transfusion hepatitis in Venezuela. *Rev. Panam. Salud. Pública*, 2001, 10 (6), 382-7.

[83] Gutiérrez, C; Devesa, M; Loureiro, CL; León, G; Liprandi, F; Pujol, FH. Molecular and serological evaluation of surface antigen negative hepatitis B virus infection in blood donors from Venezuela. *J. Med. Virol*, 2004, 73 (2), 200-7.

[84] Hadler, SC; De Monzon, M; Ponzetto, A; Anzola, E; Rivero, D; Mondolfi, A; Bracho, A; Francis, DP; Gerber, MA; Thung, S; Gerin, J; Maynard, JE; Proper, H; Purcell, RH. Delta virus infection and severe hepatitis. An epidemic in the Yucpa Indians of Venezuela. *Ann. Intern. Med*, 1984, 100 (3), 339-344.

[85] Hadler, SC; Alcala de Monzon, M; Rivero, D; Perez, M; Bracho, A; Fields, H. Epidemiology and long-term consequences of hepatitis delta virus infection in the Yucpa Indians of Venezuela. *Am. J. Epidemiol*, 1992, 136 (12), 1507-1516.

[86] Hakre, S; Reyes, L; Bryan, JP; Cruess, D. Prevalence of hepatitis B virus among health care workers in Belize, Central America. *Am. J. Trop. Med. Hyg*, 1995, 53 (2), 118-122.

[87] Hannoun, C; Norder, H; Lindh, M. An aberrant genotype revealed in recombinant hepatitis B virus strains from Vietnam. *J. Gen. Virol*, 2000, 81 (Pt 9), 2267-72.

[88] Heipertz, RA Jr; Miller, TG; Kelley, CM; Delaney, WE 4th; Locarnini, SA; Isom,HC. In Vitro Study of the Effects of Precore and Lamivudine Resistant Mutations on HBV Replication. *J. Virol*, 2007, in press.

[89] Hou, J; Karayiannis, P; Waters, J; Luo, K; Liang, C; Thomas, HC. A unique insertion in the S gene of surface antigen-negative hepatitis B virus Chinese carriers. *Hepatology*, 1995, 21 (2), 273-8.

[90] Huovila, AP; Eder, AM; Fuller, SD. Hepatitis B surface antigen assembles in a post-ER, pre-Golgi compartment. *J. Cell Biol*, 1992, 118 (6), 1305-1320.

[91] Huy, TT; Ushijima, H; Sata, T; Abe, K. Genomic characterization of HBV genotype F in Bolivia: genotype F subgenotypes correlate with geographic distribution and T^{1858} variant. *Arch. Virol*, 2006, 151 (3), 589-597.

[92] Ishikawa, K; Koyama, T; Matsuda, T. Prevalence of HBV genotypes in asymptomatic carrier residents and their clinical characteristics during long-term follow-up: the relevance to changes in the HBeAg/anti-HBe system. *Hepatol. Res*, 2002, 24 (1), 1.

[93] Isogawa, M; Robek, MD; Furuichi, Y; Chisari, FV. Toll-like receptor signaling inhibits hepatitis B virus replication in vivo. *J. Virol*, 2005, 79 (11), 7269-72.

[94] Juarez-Figueroa, L; Uribe-Salas, F; Conde-Glez, C; Hernandez-Avila, M; Olamendi-Portugal, M; Uribe-Zuniga, P; Calderon, E. Low prevalence of hepatitis B markers among Mexican female sex workers. *Sex Transm. Infect*, 1998, 74 (6), 448-450.

[95] Kao, JH; Liu, CJ; Chen, DS. Hepatitis B viral genotypes and lamivudine resistance. *J. Hepatol*, 2002, 36 (2), 303-304.

[96] Kao, JH; Chen, PJ; Lai, MY; Chen, DS. Acute exacerbations of chronic hepatitis B are rarely associated with superinfection of hepatitis B virus. *Hepatology*. 2001, 34, 817-823.

[97] Kao, JH; Chen, PJ; Lai, MY; Chen, DS. Basal core promoter mutations of hepatitis B virus increase the risk of hepatocellular carcinoma in hepatitis B carriers. *Gastroenterology*, 2003, 124 (2), 327-34.

[98] Karthigesu, VD; Allison, LM; Ferguson, M; Howard CR. A hepatitis B virus variant found in the sera of immunised children induces a conformational change in the HBsAg "a" determinant. *J. Med. Virol*, 1999, 58 (4), 346-52.

[99] Kato, H; Orito, E; Sugauchi, F; Ueda, R; Gish, RG; Usuda S; Miyakawa, Y; Mizokami, M. Determination of Hepatitis B virus genotype G by polymerase chain reaction with hemi-nested primers. *J. Virol. Methods*, 2001, 98 (2), 153-9.

[100] Kato, H; Orito, E; Gish, RG; Bzowej, N; Newsom, M; Sugauchi, F; Suzuki, S; Ueda, R; Miyakawa, Y; Mizokami, M. Hepatitis B e antigen in sera from individuals infected with hepatitis B virus genotype G. *Hepatology*, 2002, 35 (4), 922-929.

[101] Kato, H; Orito, E; Gish, RG; Sugauchi, F; Suzuki, S; Ueda, R; Miyakawa, Y; Mizokami, M. Characteristics of hepatitis B virus isolates of genotype G and their phylogenetic differences from the other six genotypes (A to F). *J. Virol*, 2002, 76 (12), 6131-6137.

[102] Kidd-Ljunggren, K; Myhre, E; Bläckberg, J. Clinical and serological variation between patients infected with different Hepatitis B virus genotypes. *J. Clin. Microbiol*, 2004, 42 (12), 5837-5841.

[103] Kirschberg, O; Schuttler, C; Repp, R; Schaeffer S. A multiplex-PCR to identify hepatitis B virus genotypes A-F. *J. Clin. Virol*, 2004, 29 (1), 39-43.

[104] Knipe, DM; Howley, PM; Griffin, DE; Lamb, RA; Martin, MA; Roizman, B; Straus, SE. *Fields Virology*. 5th edition. Lugar de publicacion : Lippincott Williams and Wilkins (LWW); 2006.

[105] Kobayashi, M ; Arase, Y ; Ikeda, K ; Tsubota, A ; Suzuki, Y ; Saitoh, S ; Kobayashi, M ; Suzuki, F ; Akuta, N ; Someya, T ; Matsuda, M ; Sato, J ; Kumada, H. Clinical characteristics of patients infected with hepatitis B virus genotypes A, B and C. *J. Gastroenterol*, 2002, 37 (1), 35-9.

[106] Konomi, N; Miyoshi, C; La Fuente Zerain, C; Li, TC; Arakawa, Y; Abe, K. Epidemiology of hepatitis B, C, E and G virus infections and molecular analysis of hepatitis G virus isolates in Bolivia. *J. Clin. Microbiol*, 1999, 37 (10), 3291-3295.

[107] Kramvis, A; Kew, MC. Relationship of genotypes of hepatitis B virus to mutations, disease progression and response to antiviral therapy. *J. Viral Hepat*, 2005a, 12 (5), 456-464.

[108] Kramvis, A; Kew, M; François, G. Hepatitis B virus genotypes. *Vaccine,* 2005b, 23 (19), 2409-2423.
[109] Kreutz, C. Molecular, immunological and clinical properties of mutated hepatitis B viruses. *J. Cell Mol. Med*, 2002, 6 (1), 113-143.
[110] Kurbanov, F; Tanaka, Y; Fujiwara, K; Sugauchi, F; Mbanya, D; Zekeng, L; Ndembi, N; Ngansop, C; Kaptue, L; Miura, T; Ido, E; Hayami, M; Ichimura, H; Mizokami, M. A new subtype (subgenotype) Ac (A3) of hepatitis B virus and recombination between genotypes A and E in Cameroon. *J. Gen. Virol*, 2005, 86 (Pt 7), 2047-2056.
[111] Kuroda, S; Fujisawa, Y; Iino, S; Akahane, Y; Suzuki, H. Induction of protection level of anti-pre-S2 antibodies in humans immunized with a novel hepatitis B vaccine consisting of M (pre-S2 + S) protein particles (a third generation vaccine). *Vaccine,* 1991,9 (3), 163-169.
[112] Laperche, S; Girault, A; Beaulieu, MJ; Bouchardeau, F; Courouce, AM. Determination of hepatitis B virus subtypes by an enzyme immunoassay method using monoclonal antibodies to type-specific epitopes of HBsAg. *J. Viral Hepat*, 2001, 8 (6), 447-53.
[113] Lee, JY; Locarnini, S. Hepatitis B virus: pathogenesis, viral intermediates, and viral replication. *Clin. Liver Dis*, 2004, 8 (2), 301-320.
[114] Lee, WM. Hepatitis B infection. *N. Engl. J. Med*, 1997, 337 (24), 1733-1745.
[115] León, G; López, JL; Maio, A; García, L; Quiroz, AM. Investigation of HBV-DNA using the polymerase chain reaction (PCR) in HBsAg-negative, anti-HBc-positive Venezuelan donors. *Sangre (Barc)*, 1999, 44 (5), 342-6.
[116] León, P; Venegas, E; Bengoechea, L; Rojas, E; Lopez, JA; Elola, C; Echevarria, JM. Prevalence of infections by hepatitis B, C, D and E in Bolivia. *Rev. Panam. Salud Pública*, 1999, 5 (3), 144-151.
[117] Levicnic-Stezinar, S. Hepatitis B surface antigen escape mutant in a first time blood donor potentially missed by routine screening assay. *Clin. Lab*, 2004, 50, 49-51.
[118] Li, J ; Xu, Z ; Zheng,Y ; Johnson, DL ; Ou, JH. Regulation of hepatocyte nuclear factor 1 activity by wild-type and mutant hepatitis B virus X proteins. *J. Virol*, 2002, 76 (12), 5875-81.
[119] Lindh, M; Anderson, AS; Gusdal, A. Genotypes, nt 1858 variants, and geographic origin of hepatitis B virus. Large scale analysis using a new genotyping method. *J. Infect. Dis*, 1997, 175: 1285-1293.
[120] Lindh, M; Hannoun, C; Dhillon, AP; Norkrans, G; Horal, P. Core promoter mutations and genotypes in relation to viral replication and liver damage in East Asian hepatitis B virus carriers. *J. Infect. Dis*, 1999, 179 (4), 775-782.

[121] Lindh, M; Horal, P; Dhillon, AP; Norkrans, G. Hepatitis B virus DNA levels, precore mutations, genotypes and histological activity in chronic hepatitis B. *J. Viral. Hepat*, 2000, 7 (4), 258-267.
[122] Liu, CJ; Chen, BF; Chen, PJ; Lai, MY; Huang, WL; Kao, JH; Chen, DS. Role of hepatitis B virus precore/core promoter mutations and serum viral load on noncirrhotic hepatocellular carcinoma: a case-control study. *J. Infect. Dis*. 2006, 194(5), 594-599.
[123] Liu, HF; Sokal, E; Goubau, P. Wide variety of genotypes and geographic origins of hepatitis B virus in Belgian children. *J. Pediatr. Gastroenterol. Nutr*, 2001, 32 (3), 274-277.
[124] Liu, Y; Hussain, M; Wong, S, Yim, HJ; Lok, AS. A genotype-independent real-time PCR assay for quantification of hepatitis B virus DNA. *J. Clin. Microbiol*, 2007, 45 (2), 553-558.
[125] Livingston, SE; Simonetti, JP; McMahon, BJ; Bulkow, LR; Hurlburt, KJ; Homan, CE; Snowball, MM; Cagle, HH; Williams, JL; Chulanov, VP. Hepatitis B virus genotypes in alaska native people with hepatocellular carcinoma: preponderance of genotype f. *J. Infect. Dis*, 2007, 195 (1), 5-11.
[126] Ljunggren, KE; Patarroyo, ME; Engle, R; Purcell, RH; Gerin, JL. Viral hepatitis in Colombia: a study of the "hepatitis of the Sierra Nevada de Santa Marta". *Hepatology*, 1985, 5 (2), 299-304.
[127] Locarnini, S. Molecular virology and the development of resistant mutants: implications for therapy. *Semin. Liver Dis*, 2005, 25 (Suppl 1), 9-19.
[128] Lok, AS. Chronic hepatitis B. *N. Engl. J. Med*, 2002, 346, 1682-1683.
[129] López, JL; Mbayed, VA; Telenta, PF; González, JE; Campos, RH. 'HBe minus' mutants of hepatitis B virus. Molecular characterization and its relation to viral genotypes. *Virus Res*, 2002, 87 (1), 41-49.
[130] Lopez, JL; Mathet, VL; Oubina, JR; Campos, RH. Intrahost evolution of HBe antigen-negative hepatitis B virus genomes ascribed to the F genotype: a longitudinal 3 year retrospective study. *J. Gen. Virol*, 2007, 88 (Pt 1), 86-91.
[131] Magnius, LO; Norder, H. Subtypes, genotypes and molecular epidemiology of the hepatitis B virus as reflected by sequence variability of the S-gene. *Intervirology*, 1995, 38 (1-2), 24-34.
[132] Margolis, HS; Alter, MJ; Hadler, SC. Hepatitis B: evolving epidemiology and implications for control. *Semin. Liver Dis*, 1991, 11 (2), 84-92.
[133] Mathet, VL; Feld, M; Espinola, L; Sánchez, DO; Ruiz, V; Mandó, O; Carballal, G; Quarleri, JF; D´Mello, FD; Howard, CR; Oubiña, JR. Hepatitis B virus S gene mutants in a patient with chronic active hepatitis with circulating Anti-HBs antibodies. *J. Med. Virol*, 2003, 69 (1), 18-26.

[134] Mathet, VL; Cuestas, ML; Ruiz, V; Minassian, ML; Rivero, C; Trinks, J; Daleoso, G; León, LM; Sala, A; Libellara, B; Corach, D; Oubiña, JR. Detection of hepatitis B virus (HBV) genotype E carried -even in the presence of high titers of anti-HBs antibodies- by an Argentinean patient of African descent who had received vaccination against HBV. *J. Clin. Microbiol,* 2006, 44 (9), 3435-3439.

[135] Mathet, VL; Lopez, JL; Ruiz, V; Sanchez, DO; Carballal, G; Campos, RH; Oubina, JR. Dynamics of a hepatitis B virus e antigen minus population ascribed to genotype F during the course of a chronic infection despite the presence of anti-HBs antibodies. *Virus Res,* 2007, 123 (1), 72-85.

[136] Mayerat, C; Mantegani, A; Frei, PC. Does hepatitis B virus (HBV) genotype influence the clinical outcome of HBV infection? *J. Viral Hepat,* 1999, 6 (4), 299-304.

[137] Mazzur, S; Bastiaans, MJ; Nath, N. Hepatitis B virus (HBV) infection among children and adults in the Solomon Islands. *Am. J. Epidemiol,* 1981, 113 (5), 510-519.

[138] Mbayed, VA; López, JL; Telenta, PF; Palacios, G; Badía, I; Ferro, A; Galoppo, C; Campos, R. Distribution of hepatitis B virus genotypes in two different pediatric populations from Argentina. *J. Clin. Microbiol,* 1998, 36 (11), 3362-3365.

[139] Mbayed, VA; Barbini, L; López, JL; Campos, RH. Phylogenetic analysis of the hepatitis B virus (HBV) genotype F including Argentine isolates. *Arch. Virol,* 2001, 146 (9), 1803-1810.

[140] McCormack, GP; Clewley, JP. The application of molecular phylogenetics to the analysis of viral genome diversity and evolution. *Rev. Med. Virol,* 2002, 12 (4), 221-38.

[141] McMahon, BJ. The natural history of chronic hepatitis B virus infection. *Semin. Liver Dis,* 2004, 24 (Suppl 1), 17-21.

[142] McMillan, JS; Bowden, DS; Angus, PW; McCaughan, GW; Locarnini, SA. Mutations in the hepatitis B virus precore/core gene and core promoter in patients with severe recurrent disease following liver transplantation. *Hepatology,* 1996, 24 (6), 1371-1378.

[143] Mendez-Correa, MC; Barone, AA; Cavalheiro, NP; Tengan, FM; Guastini, C. Prevalence of hepatitis B in the sera of patients with HIV infection in São Paulo, Brazil. *Rev. Inst. Med. Trop. S. Paulo,* 2000, 42 (2), 81-85.

[144] Meuleman, P; Libbrecht, L; Wieland, S; De Vos, R; Habib, N; Kramvis, A; Roskams, T; Leroux-Roels, G. Immune suppression uncovers endogenous cytopathic effects of the hepatitis B virus. *J. Virol,* 2006, 80 (6), 2797-807.

[145] Mizokami, M; Nakano, T; Orito, E; Tanaka, Y; Sakugawa, H; Mukaide, M; Robertson, BH. Hepatitis B virus genotype assignment using restriction fragment length polymorphism patterns. *FEBS Lett*, 1999, 450, 66-71.

[146] Mizuochi T; Okada, Y; Umemori, K; Mizusawa, S; Yamaguchi K. Evaluation of 10 commercial diagnostic kits for in vitro expressed hepatitis B virus (HBV) surface antigens encoded by HBV of genotypes A to H. *J. Virol. Methods*, 2006,136 (1-2), 254-6.

[147] Moretti, F; Bolcic, F; Cassino, L; Bouzas, B; Laufer, N; Perez, H; Cahn, P; Salomón, H; Quarleri, J. Co-infección HIV-virus de hepatitis: caracterización genómica de HBV y HCV. *Acta Gastroenterol. Latinoam*, 2006, 36 (Supl. 3), S94.

[148] Moriya, T; Kuramoto, IK; Yoshizawa, H; Holland, PV. Distribution of hepatitis B virus genotypes among American blood donors determined with a Pre-S2 epitope enzyme-linked immunosorbent assay kit. *J. Clin. Microbiol*, 2002, 40 (3), 877-80.

[149] Morozov, V; Pisareva, M; Groudinin, M. Homologous recombination between different genotypes of hepatitis B virus. *Gene*, 2000, 260 (1-2), 55-65.

[150] Motta-Castro, AR; Martins, RM; Yoshida, CF; Teles, SA; Paniago, AM; Lima, KM; Gomes, SA. Hepatitis B virus infection in isolated Afro-Brazilian communities. *J. Med .Virol*, 2005, 77 (2), 188-193.

[151] Munne, MS; Vladimirsky, S; Otegui, L; Brajterman, L; Castro, R; Soto, S; Gonzalez, J. Detección de dos variantes salvajes en hepatitis B oculta en pacientes coinfectados con virus de inmunodeficiencia humana y/o hepatitis C. *Acta Grastroenterol. Latinoam*, 2006, 36 (Supl.3), S60.

[152] Murray, JM; Wieland, SF; Purcell, RH; Chisari, FV. Dynamics of hepatitis B virus clearance in chimpanzees. *Proc. Natl. Acad. Sci. U S A*, 2005, 102 (49), 17780-5.

[153] Naito, H; Hayashi, S; Abe, K. Rapid and specific genotyping system for hepatitis B virus corresponding to six major genotypes by PCR using type-specific primers. *J. Clin. Microbiol,* 2001, 39 (1), 362-364.

[154] Nakano, T; Lu, L; Hu, X; Mizokami, M; Orito, E; Shapiro, C; Hadler, S; Robertson, B. Characterization of hepatitis B virus genotypes among Yucpa Indians in Venezuela. *J. Gen. Virol*, 2001, 82 (Pt 2), 359-65.

[155] Norder, H; Hammas, B; Lee, SD; Bile, K; Courouce, AM; Mushahwar, IK; Magnius, LO. Genetic relatedness of hepatitis B viral strains of diverse geographical origin and natural variations in the primary structure of the surface antigen. *J. Gen. Virol,* 1993, 74 (Pt 7), 1341-8.

[156] Norder, H; Couroucé, AM; Magnius, LO. Complete genomes, phylogenetic relatedness, and structural proteins of six strains of the hepatitis B virus, four of which represent two new genotypes. *Virology*, 1994, 198 (2), 489-503.

[157] Norder, H; Arauz-Ruiz, P; Blitz, L; Pujol, FH; Echevarria, JM; Magnius, LO. The T(1858) variant predisposing to the precore stop mutation correlates with one of two major genotype F hepatitis B virus clades. *J. Gen. Virol*, 2003, 84 (Pt 8), 2083-7.

[158] Norder, H; Couroucé, AM; Coursaget, P; Echevarria, JM; Lee, SD; Mushahwar, IK; Robertson, BH; Locarnini, S; Magnius, LO. Genetic diversity of hepatitis B virus strains derived worldwide: genotypes, subgenotypes, and HBsAg subtypes. *Intervirology*, 2004, 47 (6), 289-309.

[159] Ogura, Y; Kurosaki, M; Asahina, Y; Enomoto, N; Maruno, F; Sato, C. Prevalence and significance of naturally occurring mutations in the surface and polymerase genes of hepatitis B virus. *J. Infect. Dis*, 1999, 180 (5), 1444-51.

[160] Okamoto, H; Yotsumoto, S; Akahane, Y; Yamanaka, T; Miyazaki, Y; Sugai, Y; Tsuda, F; Tanaka, T; Miyakawa, Y; Mayumi, M. Hepatitis B viruses with precore region defects prevail in persistently infected hosts along with seroconversion to the antibody against e antigen. *J. Virol*, 1990, 64 (3), 1298-1303.

[161] Olinger, CM; Venard, V; Njayou, M; Oyefolu, AO; Maiga, I; Kemp, AJ; Omilabu, SA; le Faou, A; Muller, CP. Phylogenetic analysis of the precore/core gene of hepatitis B virus genotypes E and A in West Africa: new subtypes, mixed infections and recombinations. *J. Gen. Virol*, 2006, 87 (Pt 5), 1163-73.

[162] Oliveira, SA; Hacker, MA; Oliveira, ML; Yoshida, CF; Telles, PR; Bastos, FI. A window of opportunity: declining rates of hepatitis B virus infection among injection drug users in Rio de Janeiro, and prospects for targeted hepatitis B vaccination. *Rev. Panam. Salud Pública*, 2005, 18 (4-5), 271-277.

[163] Ono, SK; Kato, N; Shiratori, Y; Kato, J; Goto, T; Schinazi, RF; Carrilho, FJ; Omata, M. The polymerase L528M mutation cooperates with nucleotide binding-site mutations, increasing hepatitis B virus replication and drug resistance. *J. Clin. Invest*, 2001, 107 (4), 449-55.

[164] Oon, CJ; Chen, WN; Koh, S; Lim, GK. Identification of hepatitis B surface antigen variants with alterations outside the *a* determinant in immunized Singapore infants. *J. Inf. Dis*, 1999, 179, 259-263.

[165] Osiowy, C; Giles, E. Evaluation of the INNO-LiPA HBV genotyping assay for determination of hepatitis B virus genotype. *J. Clin. Microbiol*, 2003, 41 (12), 5473-7.
[166] Osiowy, C. Detection of HBsAg mutants. *J. Med. Virol*, 2006, 78 (Suppl. 1), S48-51.
[167] Osiowy, C; Giles, E; Tanaka, Y; Mizokami, M; Minuk, GY. Molecular evolution of Hepatitis B virus over 25 years. *J. Virol.* 2006; 80(21):10307-10314.
[168] Owiredu, WK; Kramvis, A; Kew, MC. Hepatitis B virus DNA in serum of healthy black African adults positive for hepatitis B surface antibody alone: possible association with recombination between genotypes A and D. *J. Med. Virol*, 2001, 64 (4), 441-454.
[169] Ozasa, A; Tanaka, Y; Orito, E; Sugiyama, M; Kang, JH; Hige, S; Kuramitsu, T; Suzuki, K; Tanaka, E; Okada, S; Tokita, H; Asahina, Y; Inoue, K; Kakumu, S; Okanoue, T; Murawaki, Y; Hino, K; Onji, M; Yatsuhashi, H; Sakugawa, H; Miyakawa, Y; Ueda, R; Mizokami, M. Influence of genotypes and precore mutations on fulminant or chronic outcome of acute hepatitis B virus infection. *Hepatology*, 2006, 44 (2), 326-334.
[170] Pando, MA; Berini, C; Bibini, M; Fernandez, M; Reinaga, E; Maulen, S; Marone, R; Biglione, M; Montano, SM; Bautista, CT; Weissenbacher, M; Sanchez, JL; Avila, MM. Prevalence of HIV and other sexually transmitted infections among female commercial sex workers in Argentina. *Am. J. Trop. Med. Hyg*, 2006, 74 (2), 233-238.
[171] Pando, MA; Bautista, CT; Maulen, S; Duranti, R; Marone, R; Rey, J; Vignoles, M; Eirin, ME; Biglione, MM; Griemberg, G; Montano, SM; Carr, JK; Sanchez, JL; Avila, MM. Epidemiology of human immunodeficiency virus, viral hepatitis (B and C), treponema pallidum, and human T-cell lymphotropic I/II virus among men who have sex with men in Buenos Aires, Argentina. *Sex Transm. Dis*, 2006, 33 (5), 307-313.
[172] Pavan, MH; Aoki, FH; Monteiro, DT; Goncales, NS; Escanhoela, CA; Goncales Junior, FL. Viral hepatitis in patients infected with human immunodeficiency virus. *Braz. J. Infect. Dis*, 2003, 7 (4), 253-261.
[173] Peres, AA; Dias, EA; Chesky, M; Alvares-da-Silva, MR; Jobim, LF; Goncalves, LF; Manfro, RC. Occult hepatitis B in renal transplant patients. *Transpl. Infect. Dis*, 2005, 7 (2), 51-56.
[174] Perez, OM; Morales, W; Paniagua, M; Strannegard, O. Prevalence of antibodies to hepatitis A, B, C and E viruses in a healthy population in Leon, Nicaragua. *Am. J. Trop. Med. Hyg*, 1996, 55 (1), 17-21.

[175] Ponce, JG; Cadenas, LF; García, F; León, G; Blitz-Dorfman, L; Monsalve, F; Pujol, FH. High prevalence of hepatitis B and C markers in an indigent community in Caracas, Venezuela. *Invest. Clin*, 1994, 35 (3), 123-129.
[176] Poussin, K; Dienes, H; Sirma, H; Urban, S; Beaugrand, M; Franco, D; Schirmacher, P; Brechot, C; Paterlini Brechot, P. Expression of mutated hepatitis B virus X genes in human hepatocellular carcinomas. *Int. J. Cancer*, 1999, 80 (4), 497-505.
[177] Prange, R; Mangold, CM; Hilfrich, R; Streeck, RE. Mutational analysis of HBsAg assembly. *Intervirology,* 1995, 38 (1-2), 16-23.
[178] Prange, R; Streeck, RE. Novel transmembrane topology of the hepatitis B virus envelope proteins. *EMBO J,* 1995, 14 (2), 247-56.
[179] Protzer-Knolle, U; Naumann, U; Bartenschlager, R; Berg, T; Hopf, U; Meyer zum Buschenfelde ,KH; Neuhaus, P; Gerken, G. Hepatitis B virus with antigenically altered hepatitis B surface antigen is selected by high-dose hepatitis B immune globulin after liver transplantation. *Hepatology*, 1998, 27 (1), 254-263.
[180] Proyecto Programa Nacional de Control de Hepatitis Virales: Informe Epidemiológico N° 6 [online]. August 2006. Available from: *www.anlis.gov.ar/inei/ virolog/h epatitis*.
[181] Pujol, FH; Rodriguez, I; Martinez, N; Borberg, C; Favorov, MO; Fields, HA; Liprandi, F. Viral hepatitis serological markers among pregnant women in Caracas, Venezuela: implication for perinatal transmission of hepatitis B and C. *GEN*, 1994, 48 (1), 25-28.
[182] Pumpens, P.; Grens, E.; Nassal, M. Molecular epidemiology and immunology of hepatitis B virus infection-an update. *Intervirology*, 2002; 45 (4-6), 218-232.
[183] Quintero, A; Uzcátegui, N; Loureiro, CL; Villegas, L; Illarramendi, X; Guevara, ME; Ludert, JE; Blitz, L; Liprandi, F; Pujol, FH. Hepatitis delta virus genotypes I and III circulate associated with hepatitis B virus genotype F in Venezuela. *J. Med. Virol*, 2001, 64 (3), 356-359.
[184] Quintero, A; Martínez, D; Alarcón de Noya, B; Costagliola, A; Urbina, L; González, N; Liprandi, F; Castro De Guerra, D; Pujol, FH. Molecular epidemiology of hepatitis B virus in Afro-Venezuelan populations. *Arch. Virol*, 2002, 147 (9), 1829-1836.
[185] Raimondo, G.; Balsano, C; Craxi, A; Farinati, F.; Levrero, M.; Mondelli, M.; Pollicino, T.; Squadrito, G.; Tiribelli, C. Occult hepatitis B infection. *Dig. Liver Dis*, 2000, 32 (9), 822-826.
[186] Rehermann, B; Nascimbeni, M. Immunology of hepatitis B virus and hepatitis C virus infection. *Nat. Rev. Immunol*, 2005, 5 (3), 215-29.

[187] Remondegui, C; Echenique, G; Ceballos, S; Severich, A; Roman, R; Yave, J. Prevalencia de Hepatitis B y C en bancos de sangre. *Acta Gastroenterol. Latinoam*, 2006, 36 (Supl. 3), S100.
[188] Rezende, RE; Fonseca, BA; Ramalho, LN; Zucoloto, S; Pinho, JR; Bertolini, DA; Martinelli, AL. The precore mutation is associated with severity of liver damage in Brazilian patients with chronic hepatitis B. *J. Clin. Virol*, 2005, 32 (1), 53-9.
[189] Rodríguez, L; Collado-Mesa, F; Aragon, U; Diaz, B; Rivero, J. Hepatitis B virus exposure in human immunodeficiency virus seropositive Cuban patients. *Mem. Inst. Oswaldo Cruz*, 2000, 95 (2), 243-245.
[190] Rodriguez Lay, LL; Diaz Mendiondo, B; Aragon Rodríguez, U; Delgado, G; Ante, G; Barrios Olivera, J. Infection by hepatitis B and C viruses in high-performance athletes. *Rev. Cubana Med. Trop*, 1997, 49 (3), 222-224.
[191] Rosini, N; Mousse, D; Spada, C; Treitinger, A. Seroprevalence of HbsAg, anti-HBc and anti-HCV in Southern Brazil, 1999-2001. *Braz. J. Infect. Dis*, 2003, 7 (4), 262-267.
[192] Salom, I; Roman, S; Macaya, G; Fonseca, J; Brenes, F; Mora, C; Marten, A. Retrospective review of the prevalence of hepatitis B virus in several population groups. *Rev. Biol. Trop*, 1990, 38 (1), 83-86.
[193] Sánchez, LV; Maldonado, M; Bastidas-Ramírez, BE; Norder, H; Panduro, A. Genotypes and S-gene variability of Mexican hepatitis B virus strains. *J. Med. Virol*, 2002, 68 (1), 24-32.
[194] Sánchez, LV; Tanaka, Y; Maldonado, M; Mizokami, M; Panduro, A. Difference of hepatitis B virus genotype distribution in two groups of Mexican patients with different risk factors. High prevalence of genotype H and G. *Intervirology*, 2007, 50 (1), 9-15.
[195] Sánchez-Tapias, JM; Costa, J; Mas, A; Bruguera, M; Rodes, J. Influence of hepatitis B virus genotype on the long.-term outcome of chronic hepatitis B in western patients. *Gastroenterology*, 2002, 123 (6), 1848-1856.
[196] Schaefer, S; Glebe, D; Wend, UC; Oyunbileg, J; Gerlich, WH. Universal primers for real-time amplification of DNA from all known *Orthohepadnavirus* species. *J. Clin. Virol*, 2003, 27 (1), 30-7.
[197] Schaefer, S. Hepatitis B virus: significance of genotypes. *J. Viral Hepat*, 2005, 12 (2), 111-24.
[198] Schaefer, S. Hepatitis B virus taxonomy and hepatitis B virus genotypes. *World J. Gastroenterol*, 2007, 13 (1), 14-21.
[199] Schildgen, O; Sirma, H, Funk, A; Olotu, C; Wend, UC; Hartmann, H; Helm, M; Rockstroh, JK; Willems, WR; Will, H, Gerlich, WH. Variant of

hepatitis B virus with primary resistance to adefovir. *N. Engl. J. Med*, 2006, 354 (17), 1807-12.
[200] Schilling, R; Ijaz, S; Davidoff, M; Lee, JY; Locarnini, S; Williams, R; Nauomov, NV. Endocytosis of hepatitis B immune globulin into hepatocytes inhibits the secretion of hepatitis B virus surface antigen and virions. *J. Virol*, 2003, 77 (16), 8882-92.
[201] Schmunis, GA; Zicker, F; Pinheiro, F; Branding-Bennett D. Risk for transfusion-transmitted infectious diseases in Central and South America. *Emerg. Infect. Dis*, 1998, 4 (1), 5-11.
[202] Schmunis, GA; Zicker, F; Cruz, JR; Cuchi, P. Safety of blood supply for infectious diseases in Latin American countries, 1994-1997. *Am. J. Trop. Med. Hyg*, 2001, 65 (6), 924-930.
[203] Schories, M; Peters T; Rasenack,J. Isolation, characterization and biological significance of hepatitis B virus mutants from serum of a patient with immunologically negative HBV infection. *J. Hepatol,* 2000, 33 (5), 799-811.
[204] Schurr, TG; Sherry, ST. Mitochondrial DNA and Y chromosome diversity and the peopling of the Americas: evolutionary and demographic evidence. *Am. J. Hum. Biol*, 2004,16 (4), 420-39.
[205] Scott, RM; Snitbhan, R; Bancroft, WH; Alter, HJ; Tingpalapong, M. Experimental transmission of hepatitis B virus by semen and saliva. *J. Infect. Dis*, 1980, 142 (1), 67-71.
[206] Shields, PL; Owsianka, A; Carman, WF; Boxall, E; Hubscher, SG; Shaw, J; O´Donell, K; Elias, E; Mutimer, DJ. Selection of hepatitis B surface "escape" mutants during passive immune prophylaxis following liver transplantation: potential impact of genetic changes on polymerase protein function. *Gut*, 1999, 45 (2), 306-9.
[207] Shih, WL; Kuo, ML; Chuang, SE; Cheng, AL; Doong, SL Hepatitis B virus X protein inhibits transforming growth factor-β-induced apoptosis through the activation of phosphatidylinositol 3-kinase pathway. *J. Biol. Chem*, 2000, 275 (33), 25858-64.
[208] Shouval, D. Hepatitis B vaccines. *J. Hepatol*, 2003, 39 (Suppl 1), S70-6.
[209] Silva, C; Gonçales, NS; Pereira, JS; Escanhoela, CA; Pavan, MH; Gonçales Jr, FL. The influence of occult infection with hepatitis B virus on liver histology and response to interferon treatment in chronic hepatitis C patients. *Braz. J. Infect. Dis*, 2004, 8, (6), 431-439.
[210] Silva, CM; Costi, C; Costa, C; Michelon, C; Oravec, R; Ramos, AB; Niel, C; Rossetti, ML. Low rate of occult hepatitis B virus infection among anti-

HBc positive blood donors living in a low prevalence region in Brazil. *J. Infect*, 2005, 51 (1), 24-29.

[211] Simmonds, P; Midgley, S. Recombination in the Genesis and Evolution of Hepatitis B Virus Genotypes. *J. Virol*, 2005, 79 (24), 15467-76.

[212] Sitnik, R; Pinho, JR; Bertolini, DA; Bernardini, AP; Da Silva, LC; Carrilho, FJ. Hepatitis B virus genotypes and precore and core mutants in Brazilian patients. *J. Clin. Microbiol*, 2004, 42 (6), 2455-2460.

[213] Soares, MC; Menezes, RC; Martins, SJ; Bensabath, G. Epidemiology of hepatitis B, C and D viruses among indigenous Parakana tribe in the Eastern Brazilian Amazon Region. *Bol. Oficina Sanit. Panam*, 1994, 117 (2), 124-135.

[214] Souza, MG; Passos, AD; Machado, AA; Figueiredo, JF; Esmeraldino, LE. HIV and hepatitis B virus co-infection: prevalence and risk factors. *Rev. Soc. Bras. Med. Trop*, 2004, 37 (5), 391-395.

[215] Stuyver, L; De Gendt, S; Van Geyt, C; Zoulim, F, Fried, M; Schinazi, RF; Rossau, R. A new genotype of hepatitis B virus: complete genome and phylogenetic relatedness. *J. Gen. Virol*, 2000, 81 (Pt 1), 67-74.

[216] Sucupira, MV; Mello, FC; Santos, EA; Niel, C; Rolla, VC; Arabe, J; Gomes, SA. Patterns of hepatitis B virus infection in Brazilian human inmunodeficiency virus infected patients: high prevalence of occult infection and low frequency of lamivudine resistant mutations. *Mem. Inst. Oswaldo Cruz*, 2006, 101 (6), 655-660.

[217] Sugauchi, F; Chutaputti, A; Orito, E; Kato, H; Suzuki, S; Ueda, R; Mizokami, M. Hepatitis B virus genotypes and clinical manifestation among hepatitis B carriers in Thailand. *J. Gastroenterol. Hepatol*, 2002, 17 (6), 671-676.

[218] Sugauchi, F; Orito, E; Ichida, T; Kato, H; Sakugawa, H; Kakumu, S; Ishida, T; Chutaputti, A; Lai, CL; Ueda, R; Miyakawa, Y; Mizokami, M. Hepatitis B virus of genotype B with or without recombination with genotype C over the precore region plus the core gene. *J. Virol*, 2002, 76 (12), 5985-92

[219] Suwannakarn, K; Tangkijvanich, P; Theamboonlers, A; Abe, K; Poovorawan, Y. A novel recombinant of Hepatitis B virus genotypes G and C isolated from a Thai patient with hepatocellular carcinoma *J. Gen. Virol*, 2005, 86 (Pt 11), 3027-30.

[220] Takahashi, K; Brotman, B; Usuda, S; Mishiro, S; Prince, AM. Full-genome sequence analyses of hepatitis B virus (HBV) strains recovered from chimpanzees infected in the wild: implications for an origin of HBV. *Virology*, 2000, 267 (1), 58-64.

[221] Takahashi, K; Ohta, Y; Kanai, K; Akahane, Y; Iwasa, Y; Hino, K; Ohno, N; Yoshizawa, H; Mishiro S. Clinical implications of mutations C-to-T1653 and T-to-C/A/G1753 of hepatitis B virus genotype C genome in chronic liver disease. *Arch. Virol*, 1999, 144 (7) 1299-1308.

[222] Tanaka, J. Hepatitis B epidemiology in Latin America. *Vaccine,* 2000, 18 (Suppl 1), S17-9.

[223] Telenta, PF; Poggio, GP; López, JL; Gonzalez, J; Lemberg, A; Campos, RH. Increased prevalence of genotype F hepatitis B virus isolates in Buenos Aires, Argentina. *J. Clin. Microbiol*, 1997, 35 (7), 1873-5.

[224] Teles, SA; Martins, RM; Vanderborght B; Stuyver L; Gaspar, AM; Yoshida, CF. Hepatitis B virus: genotypes and subtypes in Brazilian hemodialysis patients. *Artif. Organs*, 1999, 23 (12), 1074-8.

[225] Teles, SA; Martins, RM; Gomes, SA; Gaspar, AM; Araujo, NM; Souza, KP; Carneiro, MA; Yoshida, CF. Hepatitis B virus transmission in Brazilian hemodialysis units: serological and molecular follow-up. *J. Med. Virol*, 2002, 68 (1), 41-9.

[226] Thakur, V; Kazim, SN; Guptan, RC; Hasnain, SE; Bartholomeusz, A; Malhotra, V; Sarin, SK. Transmission of G145R mutant of HBV to an unrelated contact. *J. Med. Virol*, 2005, 76 (1), 40-6.

[227] Tillmann, HL. Antiviral therapy and resistance with hepatitis B virus infection. *World J. Gastroenterol,* 2007, 13 (1), 125-40.

[228] Toro, C; Jiménez, V; Rodríguez, C; Del Romero, J; Rodes, B; Holguin, A; Alvarez, P; Garcia-Campello, M; Gomez-Hernando, C; Guelar, A; Sheldon, J; de Mendoza, C; Simon, A; Soriano, V. Molecular and epidemiological characteristics of blood-borne virus infection among recent immigrants in Spain. *J. Med. Virol*, 2006, 78 (12), 1599-608.

[229] Torres-Baranda, R; Bastidas-Ramirez, BE; Maldonado-Gonzalez, M; Sanchez-Orozco, LV; Vazquez-Vals, E; Rodriguez-Noriega, E; Panduro, A. Occult hepatitis B in Mexican patients with HIV, an analysis using nested polymerase chain reaction. *Ann. Hepatol*, 2006, 5 (1), 34-40.

[230] Torres, JR; Mondolfi, A. Protracted outbreak of severe delta hepatitis: experience in an isolated Amerindian population of the Upper Orinoco basin. *Rev. Infect. Dis*, 1991, 13 (1), 52-5.

[231] Torres, JR. Hepatitis B and hepatitis delta virus infection in South America. *Gut*, 1996, 38 (Suppl 2), S48-55.

[232] Torresi, J.The virological and clinical significance of mutations in the overlapping envelope and polymerase genes of hepatitis B virus. *J. Clin. Virol*, 2002, 25 (2), 97-106.

[233] Tran, A; Kremsdorf, D; Capel, F; Housset, C; Dauguet, C, Petit, MA; and Brechot, C. Emergence of and takeover by hepatitis B virus (HBV) with rearrangements in the pre-S/S and pre-C/C genes during chronic HBV infection. *J. Virol*, 1991, 65 (7), 3566-74.

[234] Treitinger, A; Spada, C; Silva, EL; Miranda, AF; Oliveira, OV; Silveira, MV; Verdi, JC; Abdalla, DS. Prevalence of Serologic Markers of HBV and HCV Infection in HIV-1 Seropositive Patients in Florianopolis, Brazil. *Braz. J. Infect. Dis*, 1999, 3 (1), 1-5.

[235] Trinks, J; Cuestas, ML; Mathet, VL; Minassian, ML; Rivero, CW; Ruiz, V; Rossi, D; Rey, J; Weissenbacher, M; Martinez Peralta, L; Oubiña, JR. Estudio de la prevalencia genotípica del virus de la hepatitis B entre usuarios de drogas inyectables (UDIs) de la ciudad autónoma de Buenos Aires. *Medicina*, 2006, 66 (Suppl 1), 736.

[236] Usuda, S; Okamoto, H; Iwanari, H, Baba, K, Tsuda, F; Miyakawa,Y; Mayumi, M. Serological detection of hepatitis B virus genotypes by ELISA with monoclonal antibodies to type-specific epitopes in the Pre-S2- region product . *J. Virol. Methods*, 1999, 80 (1), 97-112.

[237] Vasquez, S; Cabezas, C; Garcia, B; Torres, R; Larrabure, G; Suarez, M; Lucen, A; Pernaz, G; Gonzales, L; Miranda, G; Davalos, E; Galarza, C; Camasca, N; Jara, R. Prevalence of HBsAg and anti-HBs carriers in pregnant women who reside in different endemic areas located in central-southern departments of Peru. *Rev. Gastroenterol. Perú*, 1999, 19 (2), 110-115.

[238] Velasco, M; Etcheverry, R. Australian antigen in various ethnic groups in Chile. *Rev. Med. Chil*, 1972, 100 (11), 1328-31.

[239] Velasco, M; Gonzalez-Ceron, M; de la Fuente, C; Ruiz, A; Donoso, S; Katz, R. Clinical and pathological study of asymptomatic HBsAg carriers in Chile. *Gut*, 1978, 19 (6), 569-71.

[240] Viana, S; Paraná, R; Moreira, RC; Compri, AP; Macedo, V. High prevalence of hepatitis B virus and hepatitis D virus in the western Brazilian Amazon. *Am. J. Trop. Med. Hyg*, 2005, 73 (4), 808-14.

[241] Vieth, S; Manegold, C; Drosten, C; Nippraschk, T; Gunther, S. Sequence and phylogenetic analysis of hepatitis B virus genotype G isolated in Germany. *Virus Genes*, 2002, 24 (2), 153-6.

[242] Vignoles, M; Avila, MM; Osimani, ML; de los Angeles Pando, M; Rossi, D; Sheppard, H; Sosa-Estani, S; Benetucci, J, Maulen, S; Chiparelli, H; Russi, J; Sanchez, JL; Montano, SM; Martinez-Peralta, L; Weissenbacher, M. HIV seroincidence estimates among at-risk populations in Buenos Aires and Montevideo: use of the serologic testing algorithm for

recent HIV seroconversion. *J. Acquir. Immune Defic. Syndr*, 2006, 42 (4), 494-500.
[243] Vinelli, E; Lorenzana, I. Transfusion-transmitted infections in multi-transfused patients in Honduras. *J. Clin. Virol*, 2005, 34 (Suppl 2), S53-60.
[244] Vivekanandan, P; Abraham, P; Sridharan, G; Chandy, G; Daniel, D; Raghuraman, S; Daniel, HD; Subramaniam, T. Distribution of hepatitis B virus genotypes in blood donors and chronically infected patients in a tertiary care hospital in southern India. *Clin. Infect. Dis*, 2004, 38 (9), e81-6.
[245] Wai, CT; Chu, CJ; Hussain, M; Lok, AS. HBV genotype B is associated with better response to interferon therapy in HBeAg(+) chronic hepatitis than genotype C. *Hepatology*, 2002, 36 (6), 1425-30.
[246] Wai, CT; Fontana, RJ. Clinical significance of hepatitis B virus genotypes, variants, and mutants. *Clin. Liver Dis*, 2004, 8 (2), 321-52.
[247] Wang, Z; Liu, Z; Zeng, G; Wen, S; Qi, Y; Ma, S; Naoumov, NV; Hou, J. A new intertype recombinant between genotypes C and D of hepatitis B virus identified in China. *J. Gen. Virol*, 2005, 86 (Pt 4), 985-90.
[248] Weber, B; Dengler, T; Berger, A; Doerr, HW; Rabenau, H. Evaluation of two automated assays for hepatitis B virus surface antigen (HBsAg) detection: IMMULITE HBsAg and IMMULITE 2000 HBsAg. *J. Clin. Microbiol*, 2003, 41 (1), 135-43.
[249] Weber, B. Genetic variability of the S gene of hepatitis B virus: clinical and diagnostic impact. *J. Clin. Virol*, 2005, 32 (2), 102-12.
[250] Weber, B. Diagnostic impact of the genetic variability of the hepatitis B virus surface antigen gene. *J. Med. Virol*, 2006, 78 (Suppl 1), S59-S65.
[251] Weinberger, KM; Zoulek, G; Bauer, T; Böhm, S; Jilg, W. A novel deletion mutant of hepatitis B vírus surface antigen. *J. Med. Virol,* 1999, 58 (2), 105-10.
[252] Weissenbacher, M; Rossi, D; Radulich, G; Sosa-Estani, S; Vila, M; Vivas, E; Avila, MM; Cuchi, P; Rey, J; Peralta, LM. High seroprevalence of bloodborne viruses among street-recruited injection drug users from Buenos Aires, Argentina. *Clin. Infect. Dis*, 2003, 37 (Suppl 5), S348-52.
[253] Westland, C; Delaney, W 4th; Yang, H; Chen, SS; Marcellin, P; Hadziyannis, S; Gish, R; Fry, J; Brosgart, C; Gibbs, C; Miller, M; Xiong, S. Hepatitis B virus genotypes and virologic response in 694 patients in phase III studies of adefovir dipivoxil. *Gastroenterology*, 2003, 125 (1), 107-16.
[254] Wilson, JN; Nokes, DJ; Carman, WF. The predicted pattern of emergence of vaccine-resistant heaptitis B: a cause for concern? *Vaccine*, 1999, 17 (7-8), 973-8.

[255] World Health Organization. Hepatitis B: World Health Organization Fact Sheet 204 [online]. October 2000. Available from: www.who.int/mediacentre/factsheets/fs204/en.

[256] Yamamoto K; Horikita, M; Tsuda, F; Itoh, K; Akahane, Y; Yotsumoto, S; Okamoto, H; Miyakawa, Y; Mayumi, M. Naturally occurring escape mutants of hepatitis B virus with various mutations in the S gene in carriers seropositive for antibody to hepatitis B surface antigen. *J. Virol*, 1994, 68 (4), 2671-6.

[257] Yang, J; Xing, K; Deng, R; Wang, J; Wang, X. Identification of Hepatitis B virus putative intergenotype recombinants by using fragment typing. *J. Gen. Virol*, 2006, 87 (Pt 8), 2203-15.

[258] Yim, HJ; Lok, AS. Natural history of chronic hepatitis B virus infection: what we knew in 1981 and what we know in 2005. *Hepatology*, 2006, 43 (2 Suppl 1), S173-81.

[259] Yuasa, R; Takahashi, K; Dien, BV; Binh, NH; Morishita, T; Sato, K; Yamamoto, N; Isomura, S; Yoshioka, K; Ishikawa, T; Mishiro, S; Kakumu, S. Properties of hepatitis B virus genome recovered from Vietnamese patients with fulminant hepatitis in comparison with those of acute hepatitis. *J. Med. Virol*, 2000, 61 (1), 23-8.

[260] Zanetti, AR; Tanzi, E; Manzillo, G; Maio, G; Sbreglia, C; Caporaso, N; Thomas, H; Zuckerman, AJ. Hepatitis B variant in Europe. *Lancet*, 1988, 332 (8620), 1132-1133.

[261] Zeng, GB; Wen, SJ; Wang, ZH; Yan, L; Sun, J; Hou, JL. A novel hepatitis B virus genotyping system by using restriction fragment length polymorphism patterns of S gene amplicons. *World J. Gastroenterol*, 2004, 10 (21), 3132-6.

[262] Zollner, B; Petersen, J; Schroter, M; Laufs, R; Schoder, V; Feucht, HH. 20-fold increase in risk of lamivudine resistance in hepatitis B virus subtype *adw*. *Lancet*, 2001, 357 (9260), 934-5.

[263] Zollner, B; Petersen, J; Schafer, P; Schroter, M; Laufs, R; Sterneck, M; Feucht, HH. Subtype-dependent response of hepatitis B virus during the early phase of lamivudine treatment. *Clin. Infect. Dis*, 2002, 34 (9), 1273-7.

[264] Zollner, B; Petersen, J; Puchhammer-Stockl, E; Kletzmayr, J; Sterneck, M; Fischer, L; Schroter, M; Laufs, R; Feucht, HH. Viral features of lamivudine resistant hepatitis B genotypes A and D. *Hepatology*, 2004, 39 (1), 42-50.

[265] Zoulim, F. Therapy of chronic hepatitis B virus infection: inhibition of the viral polymerase and other antiviral strategies. *Antiviral. Res*, 1999, 44 (1), 1-30.

Index

A

AA, 79, 84, 90, 93, 97
aberrant, 86
AC, 85
academic, 76
access, 34, 44
accidental, 31
accounting, 76
accuracy, 58
acid, 3, 6, 8, 9, 17, 21, 47, 55, 58, 59, 60, 61, 62, 63, 64, 66, 70, 73
acidic, 4, 5, 7, 9, 21, 25, 59, 60, 63, 64, 68, 73
acidity, 56
activation, 67, 96
active site, 70
acute, 6, 10, 11, 12, 37, 38, 39, 40, 41, 46, 49, 51, 59, 66, 80, 81, 85, 93, 101
acute infection, 6, 10, 12
AD, 82, 97
Adams, 83
addiction, 46
adefovir, 28, 69, 96, 100
administration, 15, 69
adolescents, 75
adult, 10
adults, 10, 51, 75, 90, 93
AF, 18, 79, 99
Africa, 2, 16, 19, 31, 32, 48, 49, 79, 92
African continent, 51, 63

age, 2, 10, 27, 34, 36, 37, 38, 39, 43
AIDS, 46
AJ, 80, 92, 101
AL, 95, 96
alanine, 11
alanine aminotransferase, 11
Alaska, 19
alcohol, 11
alcohol consumption, 11
algorithm, 99
alleles, 56
alpha, 69
ALT, 11, 26, 27
alternative, 29, 32
alternative hypothesis, 32
AM, 82, 83, 85, 86, 88, 91, 92, 97, 98
Amazon, 2, 33, 35, 37, 38, 40, 41, 42, 49, 80, 81, 83, 97, 99
Amazon River, 40
Amazonian, 78
amino, 3, 4, 5, 7, 9, 16, 21, 25, 47, 51, 55, 56, 59, 60, 61, 62, 63, 64, 66, 68, 70, 73
amino acid, 3, 4, 5, 7, 9, 16, 21, 25, 47, 51, 55, 56, 59, 60, 61, 62, 63, 64, 66, 68, 70, 73
amino acids, 16, 51, 55, 56, 60
analog, 69
animals, 31
Antibodies, 16, 58
antibody, 6, 23, 51, 58, 80, 92, 93, 101

Index

antigen, vii, 1, 3, 6, 7, 11, 29, 36, 59, 61, 64, 78, 79, 81, 82, 83, 84, 85, 86, 87, 88, 89, 90, 91, 92, 94, 96, 99, 100, 101
antigenicity, 8, 59, 73
antisera, 16
antiviral, vii, 3, 5, 12, 26, 27, 28, 29, 53, 55, 69, 70, 71, 72, 73, 74, 87, 101
antiviral drugs, 69, 70, 71
antiviral therapy, vii, 5, 29, 53, 55, 71, 72, 73, 74, 87
AP, 68, 86, 88, 89, 97, 99
apoptosis, 96
application, 90
AR, 91, 101
Argentina, vii, viii, 19, 34, 43, 44, 45, 50, 51, 61, 62, 63, 66, 67, 68, 69, 72, 84, 85, 90, 93, 98, 100
AS, 82, 88, 89, 100, 101
Asia, 2, 16, 26, 31, 32, 44
Asian, vii, viii, 25, 45, 49, 68, 88
Asian countries, vii, 68
aspartate, 11
assignment, 16, 21, 22, 48, 91
associations, 47, 51
asymptomatic, 10, 51, 60, 80, 86, 99
ATF, 68
athletes, 35, 95
Australia, 2, 19, 82
averaging, 6

B

B immune globulin, vii, 81, 94, 96
Bangladesh, 19, 46
banks, 50
Barbados, 34
base pair, 15, 17, 65
B-cell, vii, 55, 57, 67
behavior, 75
Belgium, 22, 45
binding, 3, 8, 9, 55, 68, 70, 73, 74, 84, 92
biochemical, 12, 26, 27, 64
biologic, 75
biological, vii, 25, 53, 96
biology, 53

black, 3, 9, 62, 79, 93
blood, 1, 8, 11, 12, 34, 35, 36, 37, 39, 40, 41, 42, 43, 44, 46, 47, 48, 49, 50, 52, 54, 55, 57, 58, 59, 60, 66, 67, 68, 78, 79, 81, 82, 84, 85, 88, 91, 96, 97, 98, 100
blood supply, 82, 96
blood transfusion, 57
blood transfusions, 57
Bolivia, 34, 42, 50, 86, 87, 88
bonds, 9
bootstrap, 50
Brazil, 33, 34, 39, 40, 45, 49, 61, 66, 67, 73, 78, 79, 80, 81, 82, 84, 90, 95, 97, 99
Brazilian, 40, 41, 49, 79, 82, 83, 91, 95, 97, 98, 99
breeding, 15
Buenos Aires, 34, 50, 51, 52, 84, 93, 98, 99, 100

C

Cameroon, 19, 88
Cancer, 94
capacity, 75
capital, 41, 43
carbohydrates, 6
carboxyl, 8
carcinogenesis, 26, 64
carcinoma, 2, 25, 26, 27, 36, 55, 68, 80, 85, 87, 89, 97
carcinomas, 94
Caribbean, iv, v, 31, 33, 34, 35, 68, 82
Caribbean countries, 35, 82
carrier, 10, 15, 57, 59, 86
catalytic, 4, 5
CD4, 5
CD8+, 5
CE, 83, 89
cell, vii, 5, 11, 25, 55, 56, 67, 93
cell death, 11
Central America, 17, 19, 33, 34, 36, 44, 45, 46, 47, 48, 66, 79, 86
childhood, 2
children, 10, 35, 37, 39, 42, 43, 57, 87, 89, 90
Chile, 34, 43, 99

chimpanzee, 14
China, 2, 19, 83, 100
Chinese, 86
cholesterol, 6
chromosome, 96
chronic, 2, 3, 6, 10, 11, 12, 15, 25, 26, 32, 33, 36, 38, 39, 40, 41, 43, 46, 47, 49, 51, 54, 55, 59, 60, 61, 64, 65, 67, 68, 69, 71, 73, 76, 81, 83, 84, 86, 89, 90, 93, 95, 96, 98, 99, 100, 101
chronic active hepatitis, 36, 55, 67, 89
circulation, viii, 48, 60, 61, 78
cirrhosis, 2, 55, 61, 65, 67
cis, 68
CL, 83, 85, 94, 97
classification, 15, 46
classified, 25, 29, 46, 54, 60
cleavage, 21, 24, 65
clinical, vii, viii, 5, 11, 12, 20, 26, 28, 54, 56, 63, 64, 69, 72, 76, 85, 86, 88, 90, 97, 98, 100
clinical trial, 69
clinical trials, 69
clinically significant, 28, 58
clinicians, 69
clinics, 75
clone, 62, 63
clones, 20, 61, 62, 66, 68
cloning, 22
clusters, 45, 50
Co, viii, 37, 91
codes, 3
coding, 55, 64, 74
codon, 4, 17, 54, 60, 64, 65, 67, 73
codons, 2, 7, 17, 55, 56
Colombia, 33, 37, 47, 89
colonial, 45
colonization, 31, 45
colors, 62
commercial, 29, 55, 56, 58, 59, 60, 61, 63, 91, 93
communities, 33, 35, 37, 38, 39, 40, 42, 48, 49, 75, 76, 80, 83, 91
community, 23, 38, 55, 94
complementarity, 24

complexity, 68
components, 68, 82
composite, 7
compounds, 3
conformational, 8, 9, 60, 70, 72, 87
conformational states, 8
Congress, iv
consensus, 8
consumption, 11
control, 20, 64, 75, 76, 89
correlation, 19, 22
Costa Rica, 19, 34, 36, 46
coverage, 75
CP, 92
CR, 79, 81, 87, 89
CREB, 68
crystal, 7, 70
crystal structure, 7, 70
CS, 83
CT, 93, 100
Cuba, 34, 35
cultural, 34, 38
cultural factors, 34, 38
cysteine, 60, 61, 79
cysteine residues, 60, 61, 79
cytokine, 11
cytokines, 10
cytotoxic, 54, 61

D

DA, 81, 95, 97
de novo, 69
death, 11, 39
deaths, 38
defects, 92
definition, 13
degradation, 3, 5
degree, 39, 43, 48, 59
delta, 78, 81, 86, 94, 98
Delta, 85
demographic, 96
dendritic cell, 10
density, 7, 51
destruction, 11, 85

detection, 11, 12, 20, 23, 29, 45, 55, 56, 58, 59, 60, 61, 63, 99, 100
diagnostic, vii, 17, 29, 54, 56, 58, 91, 100
diagnostic kits, 56, 91
Dienes, 94
direct repeats, 2
disease progression, 11, 27, 87
diseases, 55
disseminate, 49
distribution, 2, 7, 16, 19, 20, 31, 33, 34, 39, 45, 47, 49, 51, 75, 83, 86, 95
disulfide, 9, 56
divergence, 16, 17, 51
diversification, 14, 16, 31
diversity, iv, 13, 14, 16, 48, 80, 83, 90, 92, 96
division, 45, 50
DNA, 1, 2, 3, 5, 6, 8, 11, 12, 13, 20, 21, 22, 24, 27, 28, 29, 39, 41, 51, 52, 57, 58, 61, 63, 64, 68, 73, 81, 85, 88, 89, 93, 95, 96
DNA polymerase, 73
DNA sequencing, 20
Dominican Republic, 33, 35
donor, 48, 60, 67, 88
donors, 1, 8, 34, 35, 36, 37, 39, 40, 41, 42, 43, 44, 46, 47, 48, 49, 50, 52, 54, 58, 59, 60, 66, 78, 79, 81, 82, 84, 85, 88, 91, 97, 100
DP, 85
drug addict, 46
drug addiction, 46
drug resistance, 54, 69, 70, 92
drug use, 19, 40, 42, 44, 92, 100
drug-resistant, 69
drugs, 28, 53, 69, 70, 71
duration, 71

E

EA, 93, 97
East Asia, 88
Eastern Europe, 2, 19
economic, 38, 39, 75
ecosystems, 83
Ecuador, 34, 39
efficacy, 51, 57, 63
EIA, 29
El Salvador, 34, 36, 46
electrical, 56
electron, 7, 83
electronic, iv
electrostatic, iv
ELISA, 99
EM, 7
emergence, 12, 13, 28, 51, 56, 57, 63, 69, 76, 100
emigration, 34, 51
encoding, 5, 16, 46, 73
endogenous, 90
endonuclease, 21, 24
entecavir, 69
envelope, 4, 6, 7, 8, 9, 23, 81, 94, 98
environment, 49
environmental, 10, 13, 26
environmental factors, 13, 26
enzymatic, 5
enzymatic activity, 5
enzyme, 3, 5, 23, 29, 88, 91
enzyme immunoassay, 29, 88
enzyme-linked immunosorbent assay, 91
epidemic, 38, 85
epidemics, 38
epidemiological, 20, 23, 34, 39, 44, 57, 58, 63, 84, 98
epidemiology, iv, 20, 41, 45, 47, 49, 50, 51, 69, 76, 79, 81, 85, 89, 94, 98
epitope, 9, 55, 57, 60, 67, 91
epitopes, 5, 16, 23, 55, 57, 62, 63, 72, 73, 82, 88, 99
ER, 8, 78, 86
ES, 78
esters, 6
ethnic groups, 36, 82, 99
ethnicity, 45
etiology, 38, 39
Europe, 2, 16, 19, 72, 101
European, vii, viii, 29, 31, 45, 48, 50, 68
European Union, 29
evidence, 2, 25, 31, 96
evolution, 13, 14, 89, 90, 93
evolutionary, 14, 20, 32, 84, 96
exogenous, 53

expert, iv
expertise, 20, 24
exposure, 23, 36, 79, 95

F

factor H, 68
failure, vii, 2, 5, 8, 10, 17, 27, 40, 54, 55, 56, 58, 71
false, 23, 57, 59
false negative, 23, 57, 59
family, 1
FD, 89
FDA, 29
females, 35, 37
fever, 40
fibrosis, 26, 51
fitness, 53, 70, 72, 73
FL, 93, 96
flexibility, 8
folding, 61
FP, 84
France, viii, 17, 19, 29, 45
fulminant hepatitis, 10, 37, 40, 48, 49, 66, 101
FV, 82, 83, 85, 86, 91

G

Gabon, 19
GC, 80
gel, 21
GenBank, 18, 47
gene, 3, 4, 8, 11, 15, 17, 20, 21, 23, 26, 28, 29, 47, 50, 54, 58, 60, 61, 64, 65, 66, 68, 71, 72, 73, 74, 79, 86, 89, 90, 92, 95, 97, 100, 101
gene expression, 64, 66
generation, 8, 15, 53, 57, 88
genes, vii, 2, 4, 54, 65, 66, 68, 72, 92, 94, 98, 99
genetic, 4, 14, 25, 45, 72, 79, 83, 96, 100
genetic alteration, 72
genetic diversity, 14, 83
genetics, iv

genome, vii, 2, 3, 11, 16, 18, 19, 20, 25, 50, 53, 54, 61, 65, 68, 72, 80, 90, 97, 98, 101
genomes, 13, 20, 25, 47, 61, 73, 89, 92
genomic, vii, 14, 15, 17, 18, 21, 25, 48, 54, 65
genomic regions, 21, 25, 65
genotype, viii, 9, 12, 13, 14, 16, 17, 18, 19, 20, 21, 22, 23, 25, 26, 27, 28, 29, 31, 32, 45, 46, 47, 48, 49, 50, 51, 59, 60, 61, 62, 63, 66, 67, 68, 72, 78, 79, 81, 82, 83, 84, 86, 87, 89, 90, 91, 92, 93, 94, 95, 97, 98, 99, 100
genotypes, vii, viii, 8, 9, 11, 14, 16, 17, 18, 19, 20, 21, 22, 23, 24, 25, 26, 27, 28, 29, 31, 32, 44, 45, 46, 47, 48, 49, 50, 51, 52, 58, 60, 63, 64, 72, 75, 79, 80, 81, 83, 84, 85, 86, 87, 88, 89, 90, 91, 92, 93, 94, 95, 97, 98, 99, 100, 101
Germany, 71, 99
Gibbs, 83, 100
glass, 11
glycoproteins, 6
glycosylated, 7, 8
glycosylation, 8, 56
goals, 76
gold, 7, 20
gold standard, 20
grants, 77
Grenada, 34, 35
grouping, 16
groups, 1, 2, 19, 20, 22, 25, 29, 35, 36, 38, 40, 41, 42, 44, 46, 47, 48, 52, 80, 82, 95, 99
Guatemala, 33, 36, 46
Guyana, 35

H

H1, 18
H_2, 18
HA, 80, 84, 94
Haiti, 33, 35
half-life, 6, 11
harbour, 5
Hawaii, 19
HBV, iv, v, vii, viii, 1, 2, 3, 4, 5, 6, 7, 8, 9, 10, 11, 12, 13, 14, 15, 16, 17, 18, 19, 20, 21,

108 Index

22, 23, 24, 25, 26, 27, 28, 29, 31, 32, 33, 34, 35, 36, 37, 38, 39, 40, 41, 42, 43, 44, 45, 46, 47, 48, 49, 50, 51, 53, 54, 55, 57, 58, 59, 60, 61, 62, 63, 64, 65, 66, 67, 68, 69, 70, 71, 72, 73, 74, 75, 76, 79, 81, 84, 85, 86, 88, 90, 91, 93, 96, 97, 98, 99, 100
HBV infection, 2, 8, 10, 11, 12, 15, 26, 27, 28, 32, 33, 34, 35, 36, 37, 38, 39, 41, 42, 44, 46, 47, 54, 55, 57, 60, 61, 66, 75, 76, 85, 90, 96, 99
HCV, 11, 12, 41, 44, 81, 91, 95, 99
HD, 100
health, vii, 2, 32, 36, 37, 40, 75, 76, 86
health care, vii, 36, 40, 75, 86
health care workers, 36, 40, 86
health problems, 2
hemisphere, 40
hemodialysis, 12, 40, 46, 47, 49, 80, 98
hemophiliacs, 47
hepatitis, iv, vii, 1, 3, 5, 6, 7, 10, 11, 12, 15, 27, 33, 36, 37, 38, 39, 40, 41, 46, 47, 48, 49, 51, 55, 56, 59, 61, 62, 63, 65, 66, 67, 68, 69, 71, 73, 76, 78, 79, 80, 81, 82, 83, 84, 85, 86, 87, 88, 89, 90, 91, 92, 93, 94, 95, 96, 97, 98, 99, 100, 101
hepatitis a, 12, 27, 36, 61, 67, 83
Hepatitis B, 1, iii, iv, 1, 3, 5, 7, 10, 12, 15, 33, 37, 51, 55, 56, 59, 62, 63, 64, 66, 67, 68, 69, 71, 73, 76, 78, 79, 80, 81, 82, 83, 84, 85, 86, 87, 88, 89, 90, 91, 92, 93, 94, 95, 96, 97, 98, 100, 101
hepatitis C, 10, 11, 40, 80, 81, 91, 94, 96
Hepatitis C virus, 79
hepatitis d, 10, 86, 98
hepatocellular, 2, 25, 26, 27, 36, 40, 55, 68, 80, 85, 87, 89, 94, 97
hepatocellular carcinoma, 2, 25, 26, 27, 36, 55, 68, 80, 85, 87, 89, 94, 97
hepatocyte, 3, 10, 11, 56, 84, 88
hepatocyte nuclear factor, 84, 88
hepatocytes, 8, 10, 96
heterogeneity, 47, 80
heterogeneous, 6, 61
high-risk, 2, 35, 40, 42, 44, 47
Hispanic, 66

Hispanic population, 66
histological, 26, 89
histology, 96
HIV, 4, 11, 32, 35, 40, 41, 44, 49, 51, 61, 70, 73, 78, 80, 81, 82, 90, 91, 93, 97, 98, 99
HIV infection, 32, 78, 90
HIV-1, 32, 35, 99
HIV-2, 32
Holland, 82, 91
homeless, 42
Honduras, 33, 36, 46, 100
Hong Kong, 19
hospital, 43, 100
hospitals, 43
host, vii, 1, 6, 7, 10, 11, 15, 20, 25, 26, 53, 54, 55, 60
hot spots, 26
HTLV, 35
human, 11, 14, 31, 32, 35, 51, 70, 76, 79, 80, 84, 93, 94, 95, 97
human immunodeficiency virus, 35, 70, 80, 84, 93, 95
humans, 1, 11, 31, 32, 66, 88
hybrid, 82
hybridization, 22
hydro, 4, 7, 8, 62
hydrophilic, 4, 7, 8, 62
hydrophilicity, 28, 56, 63
hydrophobic, 7, 8
hydrophobicity, 7
hypothesis, 31, 32, 48, 49

I

icosahedral, 6, 7
identification, 16
IDU, 52
IFN, 10, 27, 69
immigrants, 45, 98
immigration, 34, 47, 51
immune globulin, vii, 81, 94, 96
immune response, 1, 10, 15, 53, 54, 55, 61, 76
immune system, vii, 9, 10, 53, 60
immunity, 12, 57
immunization, 55

immunoassays, 8, 29, 58, 59, 60, 61, 63
immunocompromised, 11
immunodeficiency, 35, 70, 80, 84, 93, 95
immunological, 58, 61, 88
immunology, 94
implementation, 75, 76
in vitro, 28, 66, 70, 73, 83, 91
in vivo, 10, 28, 86
inactive, 10
incidence, 38, 71, 76, 85
inclusion, 57
income, 38, 40
incubation, 11, 22
incubation period, 11
India, 16, 19, 100
Indian, 80
Indians, 36, 38, 42, 60, 85, 86, 91
indigenous, vii, 17, 31, 40, 41, 42, 44, 46, 47, 48, 78, 97
Indonesia, 19
infants, 10, 56, 59, 92
infection, vii, viii, 2, 6, 8, 10, 11, 12, 15, 22, 23, 25, 26, 27, 31, 32, 33, 34, 35, 36, 37, 38, 39, 41, 42, 44, 47, 48, 55, 57, 59, 60, 61, 64, 66, 67, 68, 73, 75, 76, 78, 79, 80, 81, 82, 84, 85, 86, 88, 90, 91, 92, 93, 94, 96, 97, 98, 99, 101
infections, 10, 11, 20, 21, 22, 23, 24, 28, 32, 44, 46, 47, 48, 54, 60, 75, 76, 78, 80, 87, 88, 92, 93, 100
infectious, 6, 96
infectious disease, 96
infectious diseases, 96
inflammation, 40
inflammatory, 11
inhibition, 10, 12, 64, 101
initiation, 3, 61, 65
injection, 44, 92, 100
injury, iv, 10
inmates, 34, 78
insertion, 15, 17, 23, 56, 86
interaction, 7, 68
interactions, vii, 11, 14, 54
interferon (IFN), 10, 27, 53, 54, 69, 84, 96, 100

interpretation, 7, 34
intrinsic, 20, 76
isolation, 45
isoleucine, 70

J

Jamaica, 34
Japan, 2, 19, 25
JI, 78

K

KH, 94
kinase, 96
Korea, 19

L

LAC, v, 31, 32, 33, 34, 44, 53, 54, 58, 59, 63, 69, 71, 75, 76
lamivudine, 5, 15, 28, 61, 69, 83, 86, 97, 101
Laos, 19, 46
Latin America, iv, v, vii, viii, 26, 31, 33, 37, 45, 49, 50, 58, 68, 81, 84, 96, 98
Latin American countries, vii, 49, 50, 96
lattice, 6
LC, 81, 97
lead, 9, 10
lesions, 11
LH, 84
life cycle, 11
likelihood, 10
limitation, 20, 76
limitations, 20, 28
linguistically, 13
lipid, 6
lipids, 6, 7
lipoprotein, 6
liver, vii, 2, 10, 11, 12, 15, 26, 27, 29, 39, 40, 54, 55, 56, 64, 65, 67, 70, 81, 88, 90, 94, 95, 96, 98
liver cells, 11
liver cirrhosis, 55

liver damage, 10, 88, 95
liver disease, vii, 10, 12, 26, 39, 54, 55, 64, 65, 67, 81, 98
liver failure, 2, 10, 27, 40
liver transplant, 15, 56, 70, 90, 94, 96
liver transplantation, 56, 90, 94, 96
LM, 87, 90, 100
localization, 67
location, 45, 72
long-term, 26, 74, 86
low-level, 12
lumen, 8
LV, 95, 98
lymphocyte, 61
lysis, 11

M

magnetic, iv
major histocompatibility complex, 55
Malaysia, 19
males, 35, 37, 39
management, 85
mathematical, 57
Maya, 78
measures, 75
mechanical, iv
median, 27
Mediterranean, 16, 19, 47, 51
men, 44, 93
metropolitan area, 38, 50, 51, 83
Mexican, 34, 35, 45, 46, 59, 72, 86, 95, 98
Mexico, viii, 19, 34, 45, 46, 59, 78, 81, 85
MHC, 55, 56, 73
mice, 11
microscopy, 6
Middle East, 2, 19
migration, viii, 34, 44, 45
mitochondrial, 51, 63
ML, 62, 82, 83, 90, 92, 96, 99
models, 8, 69
molecules, 7
monoclonal, 16, 23, 29, 58, 88, 99
monoclonal antibodies, 16, 23, 58, 88, 99
monoclonal antibody, 23

mononuclear cell, 12, 81
mononuclear cells, 12, 81
monotherapy, 69
morbidity, 2, 11, 32
mortality, 2, 11, 27, 32, 37, 39, 40, 48, 75
mortality rate, 27, 37, 39, 40, 48
mothers, 55, 56
mountains, 39
mRNA, 8, 64
MS, 84, 91
mutagenesis, 28, 79
mutant, vii, 15, 28, 53, 54, 55, 56, 57, 58, 59, 60, 63, 67, 70, 71, 72, 73, 79, 88, 98, 100
mutants, vii, 5, 10, 12, 13, 14, 15, 28, 29, 51, 53, 54, 55, 56, 57, 58, 59, 60, 61, 62, 63, 65, 69, 70, 71, 72, 73, 74, 76, 79, 80, 82, 85, 89, 93, 96, 97, 100, 101
mutation, vii, 13, 54, 56, 59, 61, 64, 66, 67, 68, 70, 71, 83, 84, 92, 95
mutations, vii, 5, 11, 13, 15, 21, 28, 53, 54, 55, 56, 60, 64, 65, 66, 67, 68, 69, 71, 72, 73, 79, 82, 84, 87, 88, 89, 92, 93, 97, 98, 101
MV, 71, 97, 99

N

NA, 11, 13, 80
national, 1
natural, 1, 10, 11, 55, 72, 73, 76, 85, 90, 91
necrosis, 40
neonatal, 10
neonates, 10
nested PCR, 21, 24
neutralization, 9, 55
Nevada, 37, 89
New World, 32, 45
New York, iii, iv
New Zealand, 19
NF-κB, 68
Nicaragua, 19, 34, 36, 46, 93
Nobel Prize, 1
non-infectious, 6
non-uniform, 34
normal, 11

North America, 2, 16, 19, 34, 44, 72
NS, 93, 96
N-terminal, 59, 64, 67
nuclear, 8, 68
nucleic acid, 6, 58
nucleotide sequence, 15, 54
nucleotides, 23, 64
nucleus, 69

O

Oceania, 19
oligonucleotides, 22
oligosaccharide, 8
Oncology, 85
online, 94, 101
organization, 72, 74
orientation, 83
Orinoco, 38, 98

P

PA, 11
pairing, 2, 65
Panama, 34, 36
paradox, 6
Paraguay, 34, 42
parenteral, 1
particles, 2, 3, 6, 8, 16, 63, 73, 82, 88
passive, 55, 56, 58, 96
pathogenesis, 29, 88
pathways, 68
patients, 2, 11, 12, 13, 25, 26, 27, 28, 29, 35, 36, 38, 40, 41, 44, 46, 47, 49, 51, 55, 56, 59, 61, 64, 65, 66, 67, 68, 69, 70, 71, 72, 73, 79, 81, 82, 84, 87, 90, 93, 95, 96, 97, 98, 100, 101
PBMC, 12
PCR, 11, 20, 21, 22, 23, 24, 25, 28, 29, 68, 87, 88, 89, 91
PD, 80
pediatric, 45, 50, 90
peginterferon, 69
performance, 35, 58, 95

perinatal, 1, 10, 94
peripheral blood, 12, 81
peripheral blood mononuclear cell, 12, 81
Peru, 33, 41, 45, 49, 81, 99
PF, 82, 89, 90, 98
PG, 82
phenotype, 15
phenotypic, 66
Philippines, 19
phospholipids, 6
phylogenetic, 18, 20, 24, 48, 50, 87, 92, 97, 99
PL, 80, 96
plasma, 6, 57
play, 29, 53, 64
PM, 87
point mutation, 5, 13, 15, 60, 64, 72
political, 75
polymerase, 2, 3, 4, 5, 6, 13, 28, 69, 70, 71, 72, 73, 74, 83, 84, 87, 88, 92, 96, 98, 101
polymerase chain reaction, 87, 88, 98
polymorphism, 21, 83, 91, 101
polymorphisms, 15, 21, 66
Polynesia, 19
poor, 58
population, 2, 6, 14, 15, 20, 23, 24, 34, 35, 36, 38, 39, 40, 41, 42, 44, 48, 50, 51, 53, 56, 58, 60, 61, 66, 67, 68, 73, 75, 78, 80, 82, 84, 90, 93, 95, 98
population density, 51
population group, 39, 42, 48, 80, 95
Portugal, 86
post-translational, 65
prediction, 8
pregnant, 38, 41, 43, 46, 47, 94, 99
pregnant women, 38, 41, 43, 46, 47, 94, 99
preparation, iv
pressure, 9, 13, 14, 15, 55, 56, 58, 61, 69, 74
prevention, 75
preventive, 75
preventive programs, 75
primate, 18, 31, 32
primates, 32
priming, 3, 5
probability, 31
probe, 22

procedures, 20
production, 6, 12, 53, 58, 63, 64
prognosis, 26, 27
progressive, 51, 54, 64
promoter, 13, 15, 28, 54, 64, 65, 66, 68, 79, 82, 83, 87, 88, 89, 90
property, iv
prophylaxis, 56, 76, 81, 96
protection, 1, 9, 16, 88
protein, 1, 2, 3, 4, 5, 6, 7, 8, 9, 13, 55, 56, 59, 60, 61, 62, 63, 64, 65, 66, 67, 72, 73, 88, 96
protein function, 96
proteins, 2, 3, 4, 5, 6, 7, 8, 9, 65, 68, 82, 88, 92, 94
protocols, 25
prototype, 1
public, 32
public health, 32
Puerto Rico, 35
pyrimidine, 15

R

RA, 79, 81, 82, 86, 87
random, 14
range, 13
RC, 81, 84, 93, 97, 98, 99
reactivity, 59
reading, 2, 3, 13, 15, 53, 64, 72
real time, 28, 29
real-time, 28, 89, 95
recognition, 60
recombinant DNA, 57
recombination, 14, 21, 23, 25, 26, 52, 60, 80, 85, 88, 91, 93, 97
reduction, 51
reflection, viii
refractory, 71
refugee camps, 35
regional, 34, 41, 44
relationship, 32
relevance, 10, 27, 51, 54, 63, 86
reliability, 20
remission, 26, 27, 64
renal, 41, 93

replacement, 13, 51, 84
replication, vii, 3, 5, 11, 14, 25, 26, 53, 54, 60, 64, 66, 69, 70, 71, 72, 73, 74, 83, 86, 88, 92
repo, 89
research, 1, 20, 75, 76
researchers, 50
residues, 7, 9, 17, 60, 61, 73, 79
resistance, vii, 28, 29, 54, 69, 70, 71, 72, 73, 74, 84, 86, 92, 96, 98, 101
resolution, 12
resources, 75
restriction fragment length polymorphis (RFLP), 21, 23, 24, 91, 101
retroviruses, 5
reverse transcriptase, 4, 5, 13, 70, 72, 73
RF, 92, 97
Rio de Janeiro, 40, 49, 61, 79, 92
risk, 2, 10, 11, 28, 35, 39, 40, 42, 44, 47, 55, 57, 58, 68, 75, 76, 81, 84, 87, 95, 97, 99, 101
risk factors, 57, 76, 81, 84, 95, 97
RNA, 3, 5, 13, 64, 65
rural, 38, 42
rural communities, 42
rural population, 38

S

SA, 19, 79, 82, 84, 86, 90, 91, 92, 97, 98
saliva, 1, 79, 96
sample, 20, 23
San Salvador, 44
Sarin, 98
school, 42
scientific, 23
scientific community, 23
scientists, 1, 25
scores, 26
SD, 86, 91, 92
SE, 5, 87, 89, 96, 98
secretion, 8, 56, 61, 65, 96
semen, 1, 96
sensitivity, 20, 24, 28, 29, 58
sequencing, 20, 22, 24, 29, 47
series, 29

serologic test, 99
serological marker, 35, 39, 81, 94
serology, 57, 61
serum, 1, 6, 11, 12, 21, 23, 28, 29, 46, 61, 64, 71, 81, 82, 89, 93, 96
services, iv
settlers, 44, 45
severity, vii, 10, 67, 95
sex, 36, 44, 86, 93
sexually transmitted infections, 93
Shanghai, 83
sialic acid, 8
signal transduction, 67
signaling, 10, 86
signs, 41
similarity, 45
simulation, 11
Singapore, 59, 92
sites, 3, 8, 56, 65
slave trade, 45, 48
slaves, 48, 49
social, 34, 75
Solomon Islands, 90
Somalia, 19
South Africa, 19
South America, 2, 19, 31, 32, 33, 34, 42, 44, 45, 48, 63, 66, 96, 98
Southeast Asia, 2, 26
Spain, 19, 98
speciation, 32
species, 1, 14, 28, 31, 32, 95
specificity, 58
spectrum, 61
spheres, 6, 7
stability, 53, 56
stabilization, 65
stages, 11, 12
standards, 22
steric, 70
strain, 13, 15, 16, 46, 47, 53, 57, 59
strains, vii, viii, 14, 15, 16, 20, 23, 28, 45, 46, 47, 48, 49, 50, 51, 57, 58, 59, 60, 61, 66, 67, 69, 70, 72, 73, 80, 86, 91, 92, 95, 97
strategies, 75, 76, 101
streptavidin, 22

structural protein, 3, 64, 92
subgroups, 18, 19, 45, 46
sub-Saharan Africa, 2, 32
substitution, 15, 56, 59, 62, 63, 64, 70
suburbs, 42
Sun, 101
superiority, 69
supply, 82, 96
suppression, 90
survival, 53, 55
susceptibility, 1
Sweden, 19
symmetry, 6
symptoms, 12
synthesis, 3, 5, 12, 13, 18, 64
syphilis, 80
systems, 10, 54, 76

T

T lymphocyte, 61
Taiwan, 19, 57
takeover, 99
taxonomy, 95
T-cell, 5, 11, 55, 56, 93
technology, 57, 58
teenagers, 43
Thai, 97
Thailand, 19, 97
therapeutic, 71, 72
therapy, vii, 5, 28, 29, 53, 55, 61, 69, 71, 72, 73, 74, 75, 83, 87, 89, 98, 100
three-dimensional, 8
Tibet, 19, 82
time, 2, 10, 11, 15, 20, 23, 24, 28, 29, 35, 47, 56, 88, 89, 95
time consuming, 20, 24
TNF, 10
TNF-α, 10
tolerance, 10, 11, 82
Toll-like, 86
topology, 94
trade, 45, 48
trans, 64, 74
transcriptase, 4, 5, 13, 70, 72, 73

114 Index

transcription, 3, 5, 13, 25, 64, 65, 68
transcription factor, 68
transcription factors, 68
transcriptional, 11, 53, 68
transcripts, 64
transduction, 67
transforming growth factor, 96
transforming growth factor-β, 96
transfusion, 79, 85, 96
transfusions, 57
transition, 41
transitions, 15, 54
translation, 17, 64, 65
translational, 8, 64
translocation, 8
transmembrane, 83, 94
transmission, 1, 10, 31, 32, 34, 38, 39, 43, 47, 48, 75, 76, 82, 94, 96, 98
transplant, 15, 41, 70, 93
transplant recipients, 15
transplantation, 27, 56, 81, 90, 94, 96
triglycerides, 6
Trinidad and Tobago, 34, 35
tropical areas, 42
tryptophan, 64
TT, 86

U

uniform, 34
United States, 17
urban, 37, 38, 40
urban areas, 38, 40
urbanization, 34
Uruguay, 34, 42
users, 19, 40, 42, 44, 92, 100

V

vaccination, 17, 53, 57, 61, 72, 75, 90, 92
vaccine, vii, 1, 6, 15, 51, 55, 57, 59, 60, 63, 73, 75, 88, 100
vaccines, 51, 55, 57, 96
valine, 70

values, 18, 34, 43, 50
variability, iv, 13, 14, 15, 59, 79, 89, 95, 100
variable, 6, 10, 28, 75
variation, 9, 14, 15, 20, 23, 39, 81, 85, 87
VC, 82, 97
VD, 87
Venezuela, 33, 38, 47, 48, 60, 66, 80, 83, 85, 86, 91, 94
Victoria, 77
Vietnam, 19, 46, 86
Vietnamese, 101
viral, vii, viii, 2, 3, 5, 6, 7, 10, 11, 13, 14, 15, 16, 20, 26, 36, 38, 41, 47, 48, 51, 53, 56, 57, 59, 60, 64, 65, 66, 67, 68, 69, 70, 72, 73, 74, 76, 80, 81, 83, 86, 88, 89, 90, 91, 93, 101
viral envelope, 3, 7
viral hepatitis, 80, 81, 83, 93
virological, 13, 26, 28, 98
virology, iv, 89
virus, iv, vii, 1, 3, 5, 6, 8, 10, 11, 12, 13, 14, 16, 17, 35, 49, 53, 54, 56, 58, 62, 64, 66, 67, 69, 70, 72, 73, 78, 79, 80, 81, 82, 83, 84, 85, 86, 87, 88, 89, 90, 91, 92, 93, 94, 95, 96, 97, 98, 99, 100, 101
virus infection, 78, 80, 81, 84, 85, 86, 87, 90, 91, 92, 93, 94, 96, 97, 98, 101
virus replication, 73, 86, 92
viruses, 1, 6, 13, 32, 47, 54, 55, 60, 63, 72, 82, 83, 84, 88, 92, 93, 95, 97, 100
vulnerability, 75

W

warrants, vii, 53
Watson, 79
wells, 23
West Africa, 16, 31, 92
Western Europe, 2
wild type, 15, 56, 61, 62, 64, 73
WM, 88
women, 35, 38, 41, 43, 46, 47, 80, 94, 99
workers, 34, 36, 40, 42, 44, 86, 93
World Health Organization (WHO), 2, 101

X

X-ray, 7

Y

Y chromosome, 96
young adults, 75